Organisms have disappeared as fundamental entities from modern biology, replaced by genes and their products as the primary determinants of selected characters. This is a consequence of Darwin's theory of descent with variation and survival of fitter variants. The first part of this book (by Gerry Webster) looks critically at the conceptual structure of Darwinism and describes the limitation of the theory of evolution as a comprehensive biological theory, arguing that a theory of biological form is needed to understand the structure of organisms and their transformations as revealed in taxonomy. The second part of the book (by Brian Goodwin) explores such a theory in terms of organisms as developing and transforming dynamic systems, within which gene action is to be understood. A number of specific examples, including tetrapod limb formation and *Drosophila* development, are used to illustrate how these hierarchically organized dynamic fields undergo robust symmetry-breaking cascades to produce generic forms. These are the basic morphological structures available for evolutionary transformations, whose classification into equivalence classes provides a basis for taxonomic relationships.

Evolutionary and developmental biologists, geneticists and philosophers of science will all find this a thought-provoking book.

T0215360

FORM AND TRANSFORMATION

FORM AND TRANSFORMATION

Generative and Relational Principles in Biology

GERRY WEBSTER
University of Sussex

BRIAN GOODWIN
The Open University

CAMBRIDGE
UNIVERSITY PRESS

CAMBRIDGE UNIVERSITY PRESS
Cambridge, New York, Melbourne, Madrid, Cape Town,
Singapore, São Paulo, Delhi, Tokyo, Mexico City

Cambridge University Press
The Edinburgh Building, Cambridge CB2 8RU, UK

Published in the United States of America by Cambridge University Press, New York

www.cambridge.org
Information on this title: www.cambridge.org/9780521207430

First published 1996
First paperback edition 2011

A catalogue record for this publication is available from the British Library

Library of Congress Cataloguing in Publication data

Webster, Gerry.
Form and transformation : generative and relational principles in
biology / Gerry Webster, Brian Goodwin.
p.   cm.
Includes bibliographical references and index.
ISBN 0-521-35451-X (hbk.)
1. Morphogenesis.   2. Evolution (Biology)   I. Goodwin, Brian C.
II. Title.
QH191.W435   1996
574.4–dc20   96-15492
                                                  CIP

ISBN 978-0-521-35451-6 Hardback
ISBN 978-0-521-20743-0 Paperback

Additional resources for this publication at www.cambridge.org/9780521207430

Contents

Alle Gestalten sind ähnlich und keine gleichet der andern,
Und so deutet das Chor auf ein geheimes Gesetz.

(All forms are alike and none is like another,
So that their chorus points the way to a hidden law.)

– *Goethe, "Die Metamorphose der Pflanzen"*

Preface

In a summary of his discussion of morphology in the *Origin of Species* Darwin concludes that "On this . . . view of descent with modification, all the great facts in morphology become intelligible" (1859, p. 433). This book is a contribution to that tradition in biology which disputes this conclusion.

We argue that the theory of evolution provides only limited insight into the problem of form as regards both the causal explanation of form and the relations between forms. We suggest that what is required is the development of a specific causal–explanatory theory of form, a theory of morphogenesis in the most comprehensive sense, and that such a theory will be as fundamental to biology, if not more so, at least as the theory of evolution. We contest the current view that such a theory is merely a supplement to the theory of evolution and, consequently, that it should be couched in terms of a 'genetic programme'. The end result of such a position is the disappearance from biology of organisms, conceived as fundamental and specific kinds of entities; and, to a considerable extent, this is precisely the current situation. By contrast, we argue that organisms should be regarded as the fundamental entities of biological theory. Following scholars such as Needham, Waddington and Woodger, we argue that a satisfactory theory of morphogenesis cannot be based upon an atomistic and mechanistic view of the organism but requires the development of a more adequate 'Concept of the Organism' in which organisms are treated as specific kinds of things. We advance a field theory of morphogenesis in which organisms are conceived as entities which, by virtue of their structure, are possessed of distinctive and specific generative powers, that is, natures.

Part I, written by Gerry Webster, is concerned with conceptual problems and, in particular, with the possibility of a *scientific* explanation of morphological diversity and individual forms, which was the goal of pre-Darwinian "Rational Morphology" as reconstructed by Hans Driesch. A science of form presupposes that particular forms can be conceived as instances of natural kinds; hence, the ontological status of taxa is a central issue. Evolution theory was developed, in part, as a response to the classificatory or taxonomic problem of variation, and the theory claims that individual organisms are not only related genealogically as a matter of fact but that this is the only way they can be related. Consequently, it rejects the traditional view that species (and other) taxa are classes – hence putative natural kinds – of which individual organisms are members, and replaces it with a view of species taxa as individuals – wholes – of which individual organisms are parts. Thus, if the theory of evolution is taken as providing a general metaphysical foundation for biology, it appears that there is no possibility of a scientific explanation of morphology; particular forms can only be explained by means of historical narrative, and the diversity of forms can only be unified in terms of historical genesis. In Part I, therefore, the task is, firstly, to analyse this evolutionary view and, secondly, to determine whether some view can be developed which resembles the traditional view but is free from those defects which make an evolutionary metaphysics appear attractive. Part I outlines the traditional view and considers the dialectic between empirical classification and Darwinian theory, thereby showing how the current view of the ontological status of taxa has come into being. It then considers the current view in relation to material and theoretical practice in biology. Here it is argued that this view cannot adequately sustain taxonomic and experimental practice and provides an inadequate basis for developing an explanatory theory of form. In the final chapter of Part I, an alternative dialectic is proposed between a rational systematics, of the type mooted by Driesch, and a field theory of morphogenesis.

Part II, written by Brian Goodwin, explores the structure of a theory of biological form in terms of organisms as fields. It begins with a critical look at current theories of development which are based on Weismann's separation of organisms into a genetic essence (now called a genetic programme) and a derived soma. These are examined in relation to experimental evidence and found to be inconsistent with it. Treating organisms as unified, though hierarchically complex, dynamic fields suggests a way of handling the evidence consistently. Homology emerges from this treatment as a crucial concept relating development to taxonomy, and a general definition of homology based upon developmental dynamics is proposed that has the logical structure of an equivalence relationship as used in mathematics. This is independent of history (genealogy) so that a purely relational order begins to emerge from the study of similarities and differences of organismic form at any level of the biological hierarchy, whether between parts of one organism or between organisms belonging to different taxa, independently of their genealogies.

This theme is developed in relation to a variety of morphogenetic processes in different types of organisms – unicellulars, multicellular plants and animals – to illustrate how the morphogenetic field concept provides a systematic theory for the study of biological form and transformation as revealed in evolution. The hierarchical nature of biological taxonomies can be naturally related to a basic characteristic of morphogenetic fields: during the developmental process, complex morphology emerges from initially simple forms by a cascade of symmetry-breaking bifurcations that is hierarchically organised. From this comparative study emerges the basic concept that morphogenesis is an intrinsically robust dynamic process. Organismic morphologies are then predominantly expressions of the generic modes of this generative process. In consequence, morphological species are defined by generic forms, which are natural kinds rather than historical individuals, their status in Darwinism. The task then is to study and classify the stable modes of organisms as dynamic fields which include genes, epigenesis and environmental factors. This provides the basis for a theory of morphology and taxonomy in terms of the hierarchy of morphological equivalence classes defined by the transformation sets of morphogenetic fields and their attractors.

The final chapter is an initial exploration of the structure of a generative biology, one based upon organisms as the fundamental generative units of biological form. Here a link is made with new developments in the sciences of complexity, with its emphasis on emergent order from particular types of dynamic complexity, which is the natural context for the construction of a scientific theory of form and transformation.

Acknowledgements

The arguments presented here are the result of discussions with a considerable number of friends and colleagues on a diversity of topics over a number of years. The persons involved are too numerous to thank individually, but I must single out Ted Benton, Brian Easlea and John Maynard Smith for particular mention because they also read and commented on early drafts of parts of the present text. I am especially indebted to David Hull, who most generously read the entire manuscript and whose incisive comments forced me to rethink some muddled arguments. I must also thank Roy Bhaskar for helpful advice and general encouragement. Chapter 5 owes much to Ron Brady's fine paper on Goethe's morphological concepts, the reading of which enabled me to introduce at least a small measure of order and clarity into what had previously existed as a confused collection of partial insights and half-understood ideas. Finally, I must thank my former secretary, Carol Carlisle, and my present secretary, Tessa Ellis, for their assistance throughout the period of writing this book.

Gerry Webster

The concepts developed in this book arise from interaction and reflection with a great variety of people, extending back to my PhD studies in Edinburgh under C. H. Waddington, and before that as a student at McGill University with N. J. Berrill, both of whom knew William Bateson and D'Arcy Thompson. There is a kind of intellectual lineage here with a common theme centred on the problem of biological form, with diversity understood in terms of transformation. So history *does* have a role to play after all, and I acknowledge how fortunate I have been in the people whom I have encountered on my own developmental trajectory. I am particularly grateful to Stuart Kauffman, John Maynard Smith, Lewis Wolpert, Mae-Wan Ho, Peter Saunders and Dave Lambert for extended discussions over the years about how to make sense of biological phenomena without having organisms disappear. We don't always agree, but books are places where others can't talk back. I acknowledge also the stimulus from Scott Gilbert, who read a draft of the book and, by challenging some of our propositions, elicited a number of modifications and clarifications.

Brian Goodwin

Part I

The Problem of Form

1

Introduction: Forms and Kinds

Modern biologists . . . exhibit a positive dread of form.

Aristotle and Modern Biology

Marjorie Grene's (1974, p. 408) remark is made in the course of a discussion of Aristotle and modern biology and especially of Aristotle's concept of *eidos*. Grene is not primarily concerned with the exegesis of Aristotle, a difficult enough task in itself, but rather with an attempt to draw attention to issues in biology which she regards as important and in relation to which Aristotle's concepts remain, at least potentially, significant. Her stimulating and scholarly discussion provides a convenient means of entry to the problem of form.

While *eidos,* according to Grene, is a single, univocal concept, it is used by Aristotle in two apparently different contexts. In the first, where the sense seems to be relatively straightforward, *eidos* is contrasted with *hyle;* form as against matter. In the second, where the sense is perhaps less transparent, *eidos* is contrasted with *genos;* the concept here is usually translated as species.

Consider first *eidos* and *hyle.* Here we have a pair of concepts which can be used to analyse a variety of systems. In the traditional example of Socrates' nose, snub is the form of the matter: flesh and bone. But at another 'level', bone is the form of whatever elements compose it – earth and fire, say. As Grene notes, in this context, the concept of form functions in much the same way as the concept of organisation in modern biology. The form of an entity or process denotes its principle of organisation. Thus, systems which lend themselves to analysis in these relative terms have a double aspect. On the one hand are the 'laws' or constraints arising from the nature of the material elements of which the system is composed; on the other, the

3

'laws' or constraints that arise from the order or 'arrangement' of these elements. A form–matter analysis is, therefore, antireductive in the sense that it denies that systems of the relevant kind can be completely understood in terms of the elements of which they are composed. As Grene points out, "the higher level, though dependent on the lower, is both epistemologically and ontologically prior to it" (p. 413). To know the system is to identify the kind of system it is. To explain it is to understand the constraints on the components, which make it a system of a particular kind. Since 'matter' could, in terms of its own principles, take on many 'forms', it is the existence of this organisation which makes the system the particular kind of system that it is.

In this context we can see some evidence of that "dread of form" to which Grene refers. For example, the 'revolution' in biology consequent upon developments in molecular biology is often characterised as a triumph of 'physico-chemical reductionism'. But this is to ignore its real significance. The central concepts of code, message, information and so on, are concepts pertaining to form not matter. The analysis of metabolism in terms of feedback and regulatory circuits is an analysis of the organisation of the cell, not just of the elementary chemical reactions which occur in it. In fact, Sibatani (1985), a 'first-generation' molecular biologist, has recently argued that the revolution of molecular biology "was a structuralist rather than a physical-reductionist one as had been generally believed." Webster and Goodwin (1982) independently make a similar point. The very title of Jacob's (1974) book, *The Logic of Living Systems,* which is primarily devoted to the development of molecular biology, implies a central concern with formal aspects of biology.

While there might be argument as to the exact manner in which the preceding point is best made, I do not think there would be much argument about the point per se. Why then the pronounced tendency to characterise molecular biology solely in terms of material reduction? The answer is not difficult to find. A position which is antireductionist in the sense noted and which implicitly rejects a 'crude' materialism in favour of a concept of 'informed matter' is one which asserts that living things are specific forms of being and that biology is, therefore, an autonomous science concerned with a domain of being which has its own laws; it is in effect a 'vitalist' position in the most general sense of that term (see Driesch, 1908, 1914). In the case of many biologists, a characterisation of their theoretical beliefs in such explicit terms is apt to be received with horror or derisive denial, for vitalism is the love that dare not speak its name. Moreover, the conception of biology as an autonomous science implies the rejection of a positivist conception of the unity of science. In either case, attitudes and beliefs which are deeply entrenched and which, no doubt, have an important ideological function, involving issues of identity, power and the like, are called into question.

So much for the present on *eidos* in the context of form and matter. Let us turn now to *eidos* as contrasted with *genos* – *eidos* as species. While the phenomenal entities with which most biologists (and layfolk) deal are individual organisms, such individuals are neither all identical nor all totally dissimilar. "Common Sense" (Atran, 1990) conceives them as individuals of such and such a kind possessing underlying natures. Individuals are what they are because they are individuals of a certain kind, because they 'belong' to a certain species and consequently possess a characteristic nature. And this species is what it is because it 'belongs' to a more inclusive kind, to a certain genus. Aristotle's work can be seen as a systematic explication and development of these "common sense" notions (see Atran, 1990). Within this Aristotelian tradition or paradigm, therefore, a major concern of biology was to discover the structure of real kinds. To identify the natural kinds of living things and to determine what makes them the kinds of things they are – that is, to determine their essential as opposed to their accidental properties. The end result of such an investigation is a definition, a statement of what a thing is, and this definition has a sort of explanatory power as a consequence of the subsumption of kinds. Thus, to say that "man is a rational animal" is to 'account for' the characteristics of the human species since these 'follow from' the kind of thing a human being is, from the genus and differentia. When considered outside Aristotle's conception of the teleological structure of nature, this may not seem very impressive as an 'explanation' and we might be more inclined to think in terms of 'making intelligible' rather than 'explaining'.

This question aside, if *eidos* means 'form' and also means 'species', it would seem that for Aristotle 'organisation' and 'kind' are closely linked if not identical notions. From this perspective, biological kinds cannot be characterised in terms of unstructured clusters or aggregates, for this would be an inadequate characterisation of an organism in the same way that a 'cluster of chemical reactions' would be an inadequate characterisation of the Krebs cycle. What confers upon the Krebs cycle its qualitative identity as a particular kind of metabolic cycle is its form. Likewise, the properties of the individual organisms of primary interest to biologists seem to be of a special sort in that they must be characterised in terms of a form or pattern of organisation: a whole with its parts and an 'arrangement'. The property of 'rationality', for example, might be characterised as a specific form of behaviour in which the separate acts are so ordered as to comprise a whole which can be understood in relation to an intended goal. A property like 'animality' involves a certain form of bodily structure, a certain form of physiology, a certain form of behaviour and so on.

For Aristotle, then, the being-what-it-is of an individual is its form. For Driesch (1908), "It is *form* . . . which furnishes the foundation of all biology" (p. 17). A significant, if partial, aspect of form is the bodily structure of an individual, its morphology, and, traditionally, morphology has been the basis for the identification and classification of kinds. In this context, we must think not just of a collection of parts but of an 'arrangement'. This formal aspect of individual organisms is emphasised by Bateson (1894). As he puts it: "In the bodies of living things Heterogeneity is generally orderly and formal; it is cosmic, not chaotic. . . . anyone who has ever collected . . . animals and plants . . . knows how the eye is caught by the formal regularity of an organised being . . . contrasting with the irregularity of the ground." (pp. 19–20). It is significant that Bateson's emphasis on the formal aspects of organic beings is made in the context of a critical discussion of Darwinian theory.

As Grene observes, modern biology, at least in practice, is in broad agreement with Aristotle as regards form and matter. As she also observes, there is a parting of the ways when it comes to Aristotle's concept of kinds. This, in large measure, is due to the influence of a theory – Darwinism – which for many scholars is *the* unifying theory in biology. This theory requires that the notion of natural kind be replaced by that of historical lineage and the concept of essential nature be replaced by that of the accidental collocation of properties; in traditional terms, the concept of *forma essentialis* is replaced by that of *forma accidentalis*. It is probably not coincidental that this theory has developed historically in parallel with a positivistic conception of epistemology and ontology with its notion of the independence of coexisting properties (see Harré and Madden, 1975).

It is with the notion of *eidos* as form and kind that the first part of this book is concerned. The initial objective is to examine how the traditional view has been undermined. The second objective is to determine to what extent some view can be sustained which, if not identical with that of Aristotle and "Common Sense," has at least some affinities with it.

The discussion in this part of the book largely takes the form of a critique of Darwinism. It must be emphasised that the point of such a critique is not, as many Darwinists (and anti-Darwinists) may initially suppose, the development of some new 'theory of evolution'. In recent years, and in large measure as a result of the penetrating analysis of David Hull, there has been a significant clarification of the conceptual structure of Darwinian theory and

of the roles which *scientific* and *narrative* explanations play in it. In particular, it has been argued that the theory of evolution considered as a *scientific*
explanatory theory is concerned with the *species category*. The present discussion has nothing whatsoever to say about the theory of evolution in this
sense. Rather, it is concerned with the implications of the theory as regards
taxa, which, from a Darwinist perspective, are not natural kinds. Consequently, they are not susceptible to scientific explanation but must be explained by means of narrative. Hence, a primary focus of critical attention is
the ontological status of taxa in the context of a possible science of biological
form; the arguments of David Hull provide a means of organising this discussion. The attempt to ascertain the nature and limits of Darwinism has the
object of determining a "space" in which a scientific explanatory theory of
form might flourish, and in which a new dialectic between a theory of this
kind and a rational systematics might come into being.

The nature of such a scientific theory is plain. As Driesch (1908) observes,
a living organism does not possess its 'typical' form throughout its life; rather
the form comes into being by a process of development. "So the living form
may be called a 'genetic form' . . . and therefore *morphogenesis* is the proper
and adequate term for the science which deals with the laws of organic form
in general" (p. 20). Thus, a scientific theory of form will be a theory of
morphogenesis.

There is nothing particularly novel about this approach per se. It can be
regarded, on the one hand, as a revival of the pre-Darwinian tradition of
'Rational Morphology' and, on the other hand, as an attempt to continue the
work of those scholars of the 1930s who were concerned to develop a satisfactory 'Concept of the Organism' in opposition to the prevailing mechanistic and atomistic view which characterised the Darwinist tradition.

In effect, the first part of this book can be read as a series of footnotes to,
or glosses on, the work of Driesch, Bateson, D'Arcy Thompson, Woodger,
Waddington and Goldschmidt, among others, written in the light of more
recent considerations.

organise the diversity of beings into a unified system of kinds, it sought to reveal, in phenomena which at first sight might appear to be free and unconstrained, some kind of inherent necessity.

Driesch (1908, 1914) reconstructs the pre-Darwinian study of morphology as the beginning of an attempt to develop a science of morphology which will be comparable to the existing natural sciences but which will, at the same time, take due account of the specificity of the biological object. He contrasts pre-Darwinian 'Rational Morphology' with the Darwinist historical project:

> The old morphology had sought by means of anatomy and embryology to establish the laws, if any, which actually controlled morphological phenomena. It sought, in fact to discover what morphogenesis really was. It sought, moreover, to construct what was typical in the varieties of forms, into a system which should be not merely historically determined, but which should be intelligible from a higher and more rational standpoint. (Driesch, 1914, p. 149)

As reconstructed by Driesch, therefore, Rational Morphology addresses two distinct but related issues which are, on his view, the two problems addressed by any natural science: the nomothetic problem and the systematic problem (see Driesch, 1908). According to Driesch, the nomothetic problem, concerned with "generalities," addresses itself to establishing the link between the General and the Particular, the "typical" and the individual. The laws Driesch refers to are, firstly, those governing the process "by which the type is realised for the time being in the individual" – this is what morphogenesis really is, and, secondly, "those governing how it [the type] changes its *specificity,* if such a change, i.e., a descent, is . . . assumed" (Driesch, 1914, p. 94). Thus, this first problem is primarily a causal/explanatory problem.

The second problem, concerned with "diversities," is a systematic problem; the question of kinds and the relations between kinds. Is it possible to discover or construct an "intelligible" or "rational" system of forms? By this Driesch seems to mean an attempt to discover whether there is a principle of order in the relations between apparently diverse forms; whether there are 'laws of form'. In other words, the goal of Rational Morphology is the construction not merely of an empirical classification, though this is a necessary stage, but of a rational system which would embody an understanding of the raison d'être which lies behind or beneath the empirical classification. As Driesch puts it, an empirical taxonomy is a "classificatory *preparation* for the knowledge of . . . the rational in the forms of nature" (Driesch, 1914, p. 140). A knowledge of the "rational", of the 'laws of form', would enable us to understand that there could not exist more than a certain number of diverse forms or, alternatively, that there could exist an indefinite number whose diversity was, nevertheless, related in a lawful fashion (see Driesch, 1908). Thus the project is to determine to what extent, if at all, the *merely empirical* classification of diverse individual forms can be theoretically elucidated in such a way that it, or some aspects of it, can be shown to have a systematic, formal structure, analogous to that of certain kinds of logical system, so that the diverse forms can be *intelligibly* related. I take it that the question of rational systematics is a particular instance of the more general question of the "logic of morphology" as it is referred to in the continental tradition (see Cassirer, 1950).

Driesch attributes the concept of what is called "a type" to Cuvier and Goethe and defines it as "a sort of irreducible arrangement of different parts." However, he emphasises that "all such statements [concerning 'types'] are empirical and have their limits"; nevertheless, "it is important that they are possible" (1908, p. 248). It appears, therefore, that for Driesch 'type' simply denotes the existence of *some* degree of empirical regularity in form and, therefore, *some* indication of the existence of laws or intrinsic constraints on the possible. It should also be born in mind that Driesch is writing within the paradigm of late-nineteenth-century experimental morphology (*Entwicklungsmechanik*) where the term 'typical' seems to have been used to denote *any* forms, normal or abnormal, at any stage of the 'life cycle', which might be regarded as characteristic variants of a kind insofar as they are forms which arise from the intrinsic nature of the kind as opposed to those which are externally induced and organised (see Roux, 1905; Driesch, 1908; Churchill, 1969). Consequently, the 'typical' properties of a kind need not be universal properties in the sense that they are actually possessed by every member of the kind and possessed at all times. This notion of the 'typical' does not ignore variation but rather involves the assumption that normal, variant and 'monstrous' forms are all law governed, a view also adopted by Bateson (1894; see Webster, 1992). This position, therefore, differs in crucial respects from the essentialist notion of type criticised by Mayr and Hull. This will be considered further below.

The concepts of 'type' and 'typical' have, to an extent, fallen into disrepute, largely as a consequence of the Darwinist polemics of Mayr (1959, 1963, 1966, 1976; see below), for whom 'type' is a four-letter word. However, *some* concept of the 'typical' is central to the goal of developing a *science* of form. While the realist perspective adopted in this book does not equate laws with empirical regularities and consequently does not regard the absence of such regularities as necessarily providing grounds for the absence of laws, it remains the case that, in the absence of *any* preliminary indication of empirical pattern or regularity of form, we have no prima facie grounds to suppose that there is anything to be *scientifically* investigated and explained.

The Philosophical Perspective: Realism

The philosophical position from which Driesch writes (or claims to write) is that of Transcendental Idealism. My own position is that of Realism (see Harré, 1970, 1986; Harré and Madden, 1975; Bhaskar, 1978), where scientific practice is conceived as being concerned with kinds and causes; that is, with discovering the kinds of things that exist in the world and explaining how they act. From this perspective, the two distinct aspects of any science which Driesch characterises as the concern with "generalities" and the concern with "diversities" are intimately related.

In contrast to the Positivist ontology of events, this form of Realism proposes an ontology of things or continuants – "powerful particulars" (Harré and Madden, 1975). From this perspective, the manifest properties of entities which are available to experience in the material practice of science (Harré, 1986) – that is, a practice involving the identification and classification of particulars – as well as experimental practice, have to be understood as the realisation of dispositions which are grounded in the natures of the particulars in question. Natures are determined by structures which are "hidden" and, as such, not (immediately) accessible to experience; they have to be constructed by the speculative work of theoretical imagination. The goal of the systematic enterprise of science is the construction of a real order of discrete natural kinds or taxa; as Harré (1986) argues, if taxa are to be reliable they must mark off natural kinds. Natural kind terms thus function in two dialectically related contexts: the context of material practice and that of theoretical practice.

Bhaskar (1978) distinguishes three stages in any scientific investigation. Following Locke, a distinction is made between nominal and real essence. The nominal essence of a thing consists of the behaviour or properties the manifestation of possession of which is necessary for a thing to be identified, in material practice, as being of a certain kind. Nominal essences are therefore involved in the construction of empirical taxonomies. From a realist perspective, an empirical taxonomy represents the first, *Humean,* stage which occurs in the development of any science; that is, the descriptive characterisation of putative natural kinds in terms of empirical regularities or 'proto-laws'. As I indicated earlier, it is only on this basis that we can, in the first instance, suppose that there is something to be *scientifically* explained.

Real essences, however, are those underlying structures and generative (causal) mechanisms – imagined but not necessarily imaginary – by virtue of which a thing manifests or possesses a particular behaviour or property. The discovery of natures or real essences is, therefore, an important goal of science, for these provide explanations. This is the *Lockean* level of knowledge. Thus, according to current theory, a sample of gold has the manifest properties it does because it has a particular atomic constitution which is the real essence of gold. These properties cluster because they are the properties of a kind. The real difference between gold and silver does not lie in their manifest or observable properties but in that which serves to explain these properties, that is, in what they are. Thus we arrive at the *Leibnizian* level of knowledge, for knowledge of real essences permits the formulation of real definitions, so that natural kind terms, insofar as they are based on such definitions, group together those entities whose manifest properties have a common explanation, that is, whose nature or real essence is the same. While theoretical explanations are necessarily relative to descriptions, the reality they refer to is independent of both description and explanation. However, both taxonomic and explanatory knowledge must be conceived as defeasible. As Bhaskar (1978) puts it: "Science consists of a continuing dialectic between taxonomic and explanatory knowledge; between knowledge of what kinds of things there are and knowledge of how the things there are behave. It aims at real definitions of the things and structures of the world as well as statements of their normic behaviour" (p. 211). From a realist perspective, to classify something is to commit oneself to a particular line of inquiry into the real essence. Not all general terms stand for natural kinds because not all features of the world have a common explanation and not all sets of properties individuate one and only one kind of thing. In this sense, classification is not just a matter of convenience or convention, as Darwin (1859) sometimes implies; it is correct or incorrect. Moreover, empirical taxonomies are subject to revision in the light of explanatory theories since the theory may lead to a change in the criteria for identity consequent upon which empirical properties the theory deems essential or necessary, that is which properties the thing manifests in virtue of its nature. Hence, in the dialectic between theoretical and material practice, theory always has the last word.

From this perspective, causal laws are not statements describing empirical regularities. Rather, they are statements about how things act, or rather tend to act, and things tend to act in particular and 'typical' ways because of the kinds of things they are, because of their intrinsic natures. It is important to note, however, that it is contingent whether any particular tendency will be actualised and manifest itself in terms of overt behaviour and properties. Thus it is necessary that copper, given its intrinsic nature, be a good conductor, but it is contingent whether this 'typical' property is ever made manifest for it is contingent whether a potential difference is ever applied across a sample of it. Thus knowledge of what things are, of their real essence, enables us to predict how they will typically behave and what typical characteristics they will overtly display only if we know that certain conditions prevail. In principle, therefore, the contingent nature of empirical properties does not by itself provide grounds for the rejection of realist notions of common natures. From a realist perspective natural systems are conceived as 'open' (Bhaskar, 1978). Thus, while the presence of empirical regularities provides the prima facie grounds for supposing the existence of causal laws, that is, the possibility of formulating statements about how things of a kind will tend to act, the absence of such regularities does not necessarily provide grounds for supposing their nonexistence.

2

The Old Dialectic: Empirical Classification and Darwinian Theory

Diversity

The central problem around which this debate hinges is that of morphological diversity. Indeed, the problem of diversity is effectively coterminous with the problem of form conceived in purely empirical terms because the forms of organisms are, superficially at any rate, extraordinarily variable. However, biologists have traditionally refused to regard this diversity as irreducible and have attempted, with some degree of success, to unify the diversity by representing the variable individual forms as empirical kinds of things. As Bateson explains:

> The forms of living things are diverse. They may nevertheless be separated into Specific Groups, the members of each such group being nearly alike, while they are less like the members of any other Specific Group. [The Specific Groups may by their degrees of resemblance be arranged in Generic Groups and so on.] (1894, p. 2)

Bateson here is evidently referring to the Linnaean system, which retains some "Common Sense" and Aristotelian notions insofar as it involves a logic of classes and therefore a concern with essential or 'typical' properties. We can characterise Linnaean taxonomy as being concerned to identify and relate the kinds, considered at different levels of generality or abstraction, through which individual living beings pass and repass. Assuming that the essence of nonhuman being is extension, it seeks to analyse this extension in terms of morphological parts or elements having color, size, shape, number and spatial arrangement. Its goal is to establish the natural order of empirical similarities and differences in the domain of living beings in terms of these variables and, on this basis, to assign to individual beings names which reflect their position in a natural system. Thus kinds are conceived as empirical kinds, that is, classes, and defined in terms of empirical regularities or 'laws

of coexistence', that is, observable, constant conjunctions in space of identifiable morphological properties or parts. We can regard the concept of
'type', introduced by Cuvier, as a particular methodological principle employed in this general enterprise, one which directs attention to the invariant
spatial arrangements of parts which supposedly characterise and determine
higher, more inclusive taxonomic groups.

Empirical taxonomy is, therefore, no more than the normal practice of a
science conceived on empiricist lines – that is, the search for 'laws', conceived as empirical regularities – as conditioned by the particular nature of
its object, the realm of living beings conceived synchronically.

On this basis, species taxa and, by some, other taxa, have traditionally
been regarded as natural kinds, along with the chemical elements and compounds and geometrical figures, for, on a traditional, if somewhat minimalist,
notion of 'natural kind', the term denotes the "lawlike clustering of properties" (Carr, 1987; see also Hull, 1965, 1978). From the realist perspective
adopted here, ontologies, that is, natural kind distinctions, are theory dependent. Hence, empirical classifications can be regarded as only preliminary
and tentative attempts to distinguish natural kinds because such distinctions
can be confirmed (or corrected or disconfirmed) only in terms of an explanatory theory which grounds the empirical properties employed as distinguishing criteria (see Harré and Madden, 1975; Bhaskar, 1978; Harré, 1986).

The Logical Structure of the Linnaean Hierarchy

For the purposes of this discussion it is necessary to outline the formal,
logical structure of conventional, empirical taxonomy – the Linnaean hierarchy. This has been analysed by Buck and Hull (1966), who demonstrate
that the logic of this taxonomy is a special case of the general logic of classes
– that is, a logic concerned with substances and attributes. From this perspective, therefore, the concept of a kind is the concept of a class. The Linnaean hierarchy is a structure of nested classes (taxa) of the traditional,
Aristotelian kind whose members are individual organisms. The relations
between taxa are those generated by the grouping of lower taxa under higher
taxa. These relations are, therefore, relations of class inclusion or subsumption. The names of classes (taxa) are, in principle, defined intensionally; that
is, the definition consists of a set of intrinsic (nonrelational) properties, the
possession of which by an individual organism provides the necessary and
sufficient conditions for membership of a particular taxon. These properties
are the essential properties and the intensional definition specifies the essence
of the kind. The distribution of properties amongst individual organisms de-

termines the inclusion relations, a situation which exemplifies a standard logical principle asserting the inverse variation of extension and intension. A 'higher' taxon requiring fewer properties for membership includes more individuals, that is, has greater extension. Conversely, a 'lower' taxon, because it requires more properties for membership, has a reduced extension. In this respect, a 'higher' more inclusive taxon can be said to be 'abstract' in relation to a lower taxon in that its formation involves the 'discarding' of some of the properties which are necessary for the lower taxon. Elsewhere, Hull (1969) notes that the relevant properties are, for the most part, morphological properties. For the purposes of this discussion, therefore, a classification of the Linnaean (or any other) type is to be understood as a classification of forms. It is also important to note that although evidence from embryology may be employed, the search for a systematic order in the diversity of forms is primarily concerned with adult forms, the relatively stable termini of the morphogenetic process.

The Concept of Type

The concept of 'type' is credited to Cuvier and to Goethe by Driesch (1908; see also van der Hammen, 1981, 1986). However, since there are significant differences between the concepts of these two authors, it is more appropriate to discuss Goethe's views in the different context of Chapter 5. While there are important differences between the science of Linnaeus and that of Cuvier, in certain respects their projects are similar, not least in the empiricism which characterises their methods. Cuvier's concept of 'type' is a concept of spatial organisation not unlike the seventeenth-century botanist Tournefort's concept of 'structure': "By the structure of a plant's parts we mean the composition and arrangement of the pieces that make up its body" (quoted by Foucault, 1970).

For Cuvier, the concept of 'type' is, in the first place, a principle of classification (Cassirer, 1950) so the 'type' is not a real object but a 'model' (an instrument of thought) of invariant relations, initially constructed by abstraction from concrete individuals. By means of this 'model', organisms can be compared and related since we approach the diversity of natural forms *as if* they had been constructed on the basis of a restricted number of common plans. The idea of an invariant in the form of a common structural plan or static 'schema' is, therefore, a comparative methodological principle for finding our way through the multiplicity of forms.

In their material practice, therefore, the comparative anatomists and, subsequently, embryologists, working within this tradition concentrated their attention on those observable structures which were invariant, either totally or partially – the 'typical' structures. The task as conceived for comparative embryology is summarised by Reichert writing in 1838:

> Its aim is to distinguish during the formation of the organism, the originally given, the essence of the type and to classify what is added or altered ... during the individual developments; they reach thus ... the essential structure of the organism and demonstrate the laws that manifest themselves during embryogeny. (quoted by Russell, 1916)

The majority of the early work in comparative morphology seems to have been carried out on a intuitive basis, though E. Geoffroy St Hilaire attempted to formulate the basic principle of comparison in his "Principle of Connections" (see Russell, 1916). A logically precise explication of the "Principle of Connections" in effect forms a part of the systematic discussion of the logic of the comparative method given by Woodger (1945). Here I present a schematic summary of his arguments which sacrifices some of his logical precision for the sake of clarity and brevity.

Woodger points out that to compare two things involves setting up a one-to-one relation or correspondence between their respective parts and subsequently stating how the corresponding parts resemble or differ from each other with respect to certain sets of properties. He notes that, in morphological comparisons, the relation of correspondence (homology) between parts or elements in two different forms is established on the basis of the relational properties of the parts *not* on their intrinsic properties. Each part can be uniquely characterised or designated by its position in a particular system of relations. Consequently, in comparing two forms, parts are the same, that is, are in a relation of correspondence, if they have the same designations, that is, occupy the same positions, in a numerically distinct but otherwise identical (isomorphic) system of relations. Identical systems of relations are Bauplans, a term Woodger prefers to plan, 'type' or 'schema'. Thus biological forms are conceived as patterns or wholes and, in effect, we have a kind of empirical structuralism in which diverse forms are represented as empirical transformations, a view developed by D'Arcy Wentworth Thompson (see subsequent discussion).

From this 'typological' perspective, therefore, variation is represented as 'variation on a theme' (see Brady, 1987), where the 'theme' is the invariant system of relations or structure – the Bauplan, and the variation resides in the properties – size, shape and so forth – of the elements composing the structure. The concept of a Bauplan, is therefore, a generic concept and effectively represents the 'essence' of the kind; as such it determines a class or taxon. Woodger explains how different Bauplans can be conceived as "overlapping" and how, given the Bauplans determining a number of taxa together with the overlapping relations, the hierarchical inclusion relations between taxa are established. He points out that, on the whole, Bauplans and overlapping relations are only useful in determining and relating higher, more inclusive taxa; the best example is the phylum Chordata. At lower taxonomic levels, they play little or no part in the determination of taxa.

Essentialism

Writing from a Darwinist perspective, Mayr has characterised the classical approach to diversity which I have outlined in terms of a philosophical stance which he sees as "a major misconception" (Mayr, 1963, p. 5) having its origins in "the idealistic philosophy of Plato." He explains:

This philosophy [essentialism] when applied to the classification of organic diversity, attempts to assign the variability of nature to a fixed number of basic types at various levels. It postulates that all members of a taxon reflect the same essential nature, or in other words that they conform to the same type. This is why the essentialist ideology is also referred to as typology. Variation, consequently, is considered by the typologist as trivial and irrelevant. (1969, p. 67)

Whether, as a matter of historical fact, the biological concept of 'type' is derived from Plato's theory of Forms, as Mayr claims, is debatable (see van der Hammen, 1981). It is also debatable whether any natural historians have ever, in practice, adhered to a strict form of essentialism (see Atran, 1990). More significantly, the claim that 'typologists', from Aristotle onwards, have "considered" variation as "trivial and irrelevant" or as an "illusion" (Mayr, 1963) must be regarded, at best, as a polemical and one-sided exaggeration. Indeed, one could argue that the very nature of the "essentialist method" forces those who employ it to confront, rather than ignore, empirical variation as a taxonomic problem since essentialism is a philosophy of discontinuity. As I noted, far from treating variation as "irrelevant" or an "illusion," "typologists" – in the sense of those who subscribe to the concept of "a type" – attempt to discover invariants, 'Unity of Type', in order that variation can be represented in terms of 'variation on a theme' in the sense explicated by Woodger. However, the "essentialist" method of invariant 'types' implies that, other things being equal, forms which vary discontinuously are putatively different in kind. Hence, apparent 'variation *of* the theme' raises further problems and the 'solution' proposed or implied, which I discuss subsequently, is also employed by Darwin. From this perspective, the history of "typology" is, in fact, the history of a strenuous attempt to comprehend variation rather than to treat it as "irrelevant." Whether these attempts are successful is another matter.

However, a philosophy of discontinuity cannot serve as a means of systematically representing variation which is, or appears to be, continuous. Thus, insofar as the empirical variation of a set of forms is actually continuous in any sense, then any actual empirical classification of forms will inevitably involve some degree of misrepresentation if constructed on an essentialist basis. I presume that this must be the issue which Mayr is addressing when he speaks of "typologists" treating variation as "irrelevant" or as an "illusion."

Let us first examine the problems raised for the concept of a 'type' by discontinuity – that is, apparent 'variation *of* the theme.' Now, it is a fact that a good deal of empirical variation does not seem to respect the integrity of the supposedly 'common plan' and not all features of the 'type' are present in each supposed exemplification of it. How, then, can we retain the concept of a 'common plan' or 'Unity of Type', hence, the idea that we have a single kind, while at the same time accommodating the variation? It might be done by conceiving the 'type' as an 'Ideal Type' so that variation can be conceived as variation *on* this 'Ideal Theme'. The 'pentadactyl limb', supposedly characteristic of (exclusive to) Tetrapods – their "fundamental plan of structure" (de Beer, 1932), seems to be an example of such an 'Ideal Type'. Thus, while morphological correspondences can be established between *some* of the parts of individual instances of entities which share the name 'pentadactyl limb', the fact that some Tetrapods actually possess fewer than five digits on any limb, that some occasionally possess more and that many possess different numbers of digits on the fore and hind limbs means that the sets of parts in different individuals are not completely isomorphic. Since the variation between limbs is discontinuous in this respect, one might be tempted to wonder whether there is any such kind as *the* 'pentadactyl limb'. Is it, perhaps, merely a nominal kind? However, morphologists seem not to treat it as such. Rather, they attempt to save the notion of 'variation on a theme' by treating the variant forms as variants of a single, real kind; an 'Ideal Type'. Consequently, they continue to ask questions concerning morphological correspondence such as the one posed by Kent (1969) regarding the relations between the hand of *Rana catesbiana,* which has five digits, and the hand of *Necturus,* which has four: "Which finger is missing in Necturus?"

It will be remembered that the comparative method makes use only of the relational properties of elements in establishing morphological correspondences. Consequently, as Woodger (1945) points out, a change in the number of elements raises difficulties for comparison because it changes the relational structure of the system. These difficulties can sometimes be overcome by increasing the number of relational properties which are taken into account, but sometimes this does not suffice. In this case, recourse is made to embryological studies where the complete set of parts can sometimes be found and the basic idea of 'variation on a theme' can be saved by invoking differential growth of the parts during development. Alternatively, and more problematically, an attempt is made to find a complete set of parts by recourse to supposed ancestral forms. In either case, as Woodger observes, we might be led to the conclusion that a single element in one form corresponds to more than one element in another, thus violating the notion of correspondence as a one-to-one relation. In any event, it is evident that questions of the kind posed by Kent are, at best, difficult to answer in a determinate fashion; at worst, they are impossible since, as Woodger notes, it is only in forms with bilateral symmetry that a determinate maximum identity correspondence can be established. Bateson (1894) goes further and claims that the very asking of some questions of this kind borders on the absurd: "Which vertebra of a Pigeon, which has 15 cervical vertebrae, is homologous with the first dorsal of a Swan which has 26 cervicals?" (p. 33). Bateson also observes that the notion of a 'common plan' is invoked somewhat arbitrarily, in that there are situations regarding which questions of morphological correspondence are, in practice, never asked. He claims that in cases where individual elements exhibit no particular substantive differentiation, such as the individual segments of the gut of an earthworm, the teeth of bony fish or the leaves on a tree, no one supposes that there is any morphological correspondence between the parts of such structures. Insofar as any form of correspondence is established, this is only between the structures as wholes. We might also note that, with respect to numerical variation in forms which display rotational symmetry, questions of this kind cannot, in principle, be asked because no determinate correspondence between the parts can be established (Woodger, 1945). Thus, we have no empirical basis for treating say, two *Crocus* flowers, displaying threefold and fourfold rotational symmetry respectively, as variants of a single kind. Empirically, they are irreducibly diverse and comprise two distinct and unrelated kinds (compare Bateson, 1894, p. 60).

The fact that questions of the kind posed by Kent and cited by Bateson are asked leads one to wonder whether a methodological principle has been reified and given ontological, and even explanatory, significance. Whether, in Kantian terms, a purely 'subjective maxim' or regulative principle has been transformed into an 'objective', constitutive principle (see Cassirer, 1950) so that, instead of approaching the world *as if* it had been constructed on the basis of an invariant empirical plan or plans, morphologists assume that it has been so constructed as a matter of fact and that, somewhere, the 'Ideal Type' really exists. It is difficult to imagine the mode of existence of such an abstract object, a pure system of relations, in which, by definition, the relata lack any determinate qualities. Consequently, it is hardly surprising that, when Owen (1848) attempts a diagram of the "Vertebrate Archetype," it turns out to be just another particular.

Darwin (1859) is firmly within the 'typological' tradition, as is evident from his claim that the limbs of diverse mammals are "all constructed on the same pattern, and . . . include the same bones, in the same relative positions" (p. 415). Although, in the spirit of Comtean positivism, he dismisses the explanatory conception of 'Type' as "a mere metaphysical principle" (Darwin, 1889), he is so steeped in the tradition that he (1859, p. 416) finds it natural to transform Owen's "archetype" which, whatever Owen may have thought, can only be some kind of generic concept, into the "ancient progenitor," which is a particular. The claim concerning the sameness of empirical "pattern" is used as evidence for a genetic connection between forms and, conversely, it is claimed that "unity of type is explained by unity of descent" (1859, p. 233). On the basis of this materialist reification, Darwin and his disciples in morphology can continue employing almost exactly the same 'typological' concepts and methodology as their predecessors and, consequently, being confronted by the same problems. Now, however, instead of empirical forms being referred to an "Ideal Type," they are referred to a "Common Progenitor," and instead of constructing taxa, the task is to construct pedigrees.

The theory of "descent with modification" thus retains the 'typological' concept of variation as 'variation on a theme'. When Darwin addresses the question of apparent 'variation *of* the theme' as presented by a numerical variation in the number of elements or parts within a 'plan', he resorts to the standard 'typological' strategy outlined above; that is, he changes the 'standard' notion of a one-to-one relation of correspondence between elements into a one-to-many relation. Thus he claims that "the general pattern of an organ might become so much obscured as to be finally lost by the atrophy . . . abortion . . . soldering together . . . and doubling or multiplication" of parts (1859, p. 417). Now, these sorts of claims are relatively unproblematic if they refer to changes occurring during the embryonic development of a single individual, since the spatio-temporal continuity, hence the identity, of individual parts can, in principal, be observed. However, the claims are much more problematic if they are taken, as Darwin clearly intends them to be taken, as referring to historical changes and, consequently, as referring to the differences between individuals. In this case, the identity of parts cannot be grounded in observations of continuity; rather, it is inferred on the basis of correspondence relations. As Bateson (1894) observes, the inference involves an assumption concerning the nature of variation. The Darwinian hypothesis assumes the existence of an ancestral form, characterised by some maximum number of parts, from which the pattern characteristic of the descendents has arisen by successive diminution of parts. It is assumed, therefore, "that in Variation the individuality of each member of a Meristic Series is respected" such that parts in different individuals within an historical sequence can be identified; each part has "an individual and proper history" (p. 32). In other words, Bateson is claiming that the methodological 'atomism' which is required by the comparative method – the establishment of correspondences between individual elements – has been translated into an ontological 'atomism' whereby parts actually are independent individuals and behave as such. Bateson claims that the empirical study of variation suggests that, although this assumption sometimes appears to be justified, this is by no means always the case. He also observes, as I noted earlier noted, that 'individual homologies' are not always invoked and, in the case of forms with rotational symmetry, they cannot be invoked. Bateson claims that there has been a failure to recognise "the unity of Meristic Repetition wherever found" (p. 32). Meristic series must be conceived as patterns, not as aggregates of independent elements; hence, when they vary, they do not vary in a piecemeal fashion but as wholes.

Thus, insofar as we can suppose, on empirical grounds, that numerical variation does not respect 'individual homologies' or, insofar as we have no grounds, because of methodological limitations, for supposing that it does, we are led to the conclusion that numerical variants are, empirically, different kinds of things: 'original discontinuities' which are unrelated. Consequently, this apparently irreducible diversity cannot be unified historically; unification can only be achieved, if at all, at another level. For Bateson, this is the level of the generative mechanisms that produce the forms, a matter which will be pursued in Chapter 4.

Let me turn now to the problems posed for empirical classification by continuous variation, and, in particular, for the construction of definitions of the names of taxa. Aristotelian classes are discrete or discontinuous entities and Buck and Hull (1966) draw attention to the 'taxonomic problem of variation' in their account of the logical structure of the Linnaean system. They point out that in consequence of variation amongst the individuals which are taken to comprise the extension of the name of a taxon, taxonomists are seldom, if ever, able to provide a precise and determinate set of empirical properties which can serve as the intensional definition of the name of a taxon. Conversely, if a precise intensional definition is proposed, many individuals are, or strictly should be, excluded from the extension. If such individuals are not, in fact, excluded then we have the paradoxical notion of a class with abnormal members.

It would seem, therefore, that insofar as an Aristotelian definition of the name of a taxon *is* actually formulated, this can be accomplished only by means of idealisation in the sense that any variation that exists among the members of the taxon is discounted and, perhaps on the basis of statistical distribution of properties and the idea of a 'normal' member, a particular set of properties is designated as the essential or 'typical' properties. Hull expresses his opinion on this with characteristic forcefulness:

> Presented with a welter of diverse forms to be classified, a taxonomist can greatly simplify his task if he pretends that certain properties are 'essential' for definition. But he would have to do just that – pretend – since the names of taxa cannot be defined in terms of essential characters without falsification on a scale which should have been evident even to the most uncritical investigator with only a limited knowledge of the organisms being classified. (1965, p. 316)

Hull (1965) claims that taxonomists have, as a matter of fact, abandoned Aristotelian definitions of the names of taxa and that it is widely recognised

> that taxa names cannot be defined by sets of properties the members of which are severally necessary and jointly sufficient, for seldom is a property of any taxonomic value distributed both universally and exclusively among the members of a taxon. The properties which are used to define the names of taxa do not respect taxonomic boundaries. (p. 322)

Hull illustrates his point with a discussion of the properties actually used to define the phylum Chordata. Depending on whether the hemichordates are included or excluded, none of the defining properties are both necessary and sufficient. If Chordata is taken to include the hemichordates then, although there are a number of properties which are possessed exclusively by the chordates, none are possessed universally. On the other hand, if the hemichordates are not included in Chordata but regarded as a separate phylum, then a number of properties are now universally possessed by chordates but some properties which, in the first situation were exclusively chordate, cease to be exclusively possessed by them. The only property which is both universal and exclusive to Chordata is the possession of a notochord.

Hull concludes that "Aristotelian definition simply won't do." Rather, he argues, taxonomists actually define, and can only define, the names of taxa as "cluster concepts" employing "sets of statistically covarying properties arranged in indefinitely long disjunctive definitions.'' (p. 323). Hence, insofar as taxa can be conceived as kinds, then these are "non-standard kinds" (Hull, personal communication). However, as will be explained shortly, in his later discussions (post-1974) Hull argues that taxa are not kinds at all, whether standard or nonstandard.

What Hull seems to be claiming in his 1965 discussion is that, as a matter of fact, the members of a taxon possess few if any (intrinsic) properties which can be regarded as essential properties, where a criterion for a property being essential is its universality; I argue later in this book that, from a realist perspective, essential properties should be conceived in terms of necessity rather than universality. In the present context, if no properties are conceived as essential then, in terms of the traditional opposition, all properties must be conceived as accidental. It would appear, therefore, that we have to accept that whatever properties a given individual actually possesses, it could, in principle, simultaneously possess any other properties whatsoever. Thus, insofar as properties are conceived as independent then, again, we are invited to subscribe to a form of organismic 'atomism'. Although Hull's claims regarding essentialism which I have discussed so far are based largely on empirical considerations, they are consistent with his analysis of the ontological status of taxa as they are conceived within Darwinian theory (see later discussion). If accepted, the claims raise important questions concerning criteria for identity; I outline Hull's views on this matter subsequently and discuss them more fully in Chapter 3.

For Mayr (1963), essentialism is a "major misconception that had to be eliminated before a sound theory of evolution could be proposed." For Hull (1965) it is a conception that has resulted in "two thousand years of stasis" in taxonomy, a claim, incidentally, which is contested by Atran (1990), who argues that this is a tendentious view of history. For both Mayr and Hull, essentialism in taxonomy is a sinful and pernicious doctrine, and both quote approvingly from Popper's (1966, p. 9) well-known diatribe against the "regrettably still prevailing intellectual influence" of Aristotle and his characterisation of "the essentialist method" as "empty verbiage and barren scholasticism." Moreover, "the degree to which the various sciences have been able to make any progress depended upon the degree to which they have been able to get rid of this essentialist method." However, Hull himself (1976, 1978) accepts a form of essentialism in chemistry, and by no means all philosophers are as dismissive of essentialism as Popper. In recent years a form of essentialism has been reinstated by a number of philosophers of a broadly realist persuasion, including Putnam (1975), Bhaskar (1978), Kripke (1980) and Harré (1986). Moreover, as I will argue shortly, it is at least debatable whether the attempts by Mayr and Hull to employ Darwinian theory as the means by which essentialist notions can be eliminated from the study of morphology are successful.

Insofar as the critique of essentialism is directed to showing that diversity cannot, in practice, be systematically represented in terms of a logic of empirical classes of the traditional kind, then, I believe, the claims of the Darwinists have to be accepted. However, this very fact suggests that the Darwinist critique is, in a sense, misdirected. The real source of the 'taxonomic problem' is not essentialism per se but empiricism and a practice of classification which follows from empiricist notions; in particular the empiricist doctrine of essence and theory of concept formation and the view that kinds are to be equated with classes. In schematic terms, this doctrine claims that we start with given, concrete particulars and, by a process of comparison, abstract the 'common features', discarding the remainder. In this way we arrive at the specific and, subsequently, the generic characteristics. Thus, the 'essence' of a kind, whether it be conceived in terms of an intensional definition or a 'type' or a 'model', is nothing more than the abstracted 'common element' which serves to characterise a class.

Thus, the Aristotelian hierarchy of concepts – genus and species – as it appears in the Linnaean system is supposedly constructed by means of a process of abstraction (see van der Hammen, 1981, 1986). That is, starting with a given set of concrete individual organisms, we compare these particulars in an attempt to discover common properties. Once discovered, such properties are abstracted and used to define the name of a class (*Fritillaria meleagris,* say) whilst the remaining properties, those unique to the particulars, are discarded. The process is repeated and an order and division of forms in terms of genus and species arises which is based on the factual similarities existing in the concrete particulars. On this basis, we regard, for example, Liliaceae as the name of a discrete inclusive class (a family) within which are subsumed the discrete classes (genera) named *Fritillaria, Tulipa, Lilium* and so forth and *Fritillaria* as the name of an inclusive class within which are subsumed the discrete classes (species) named *F. meleagris, F. pontica, F. pyrenaica* and so forth.

As Cassirer (1923, p. 5) observes, the superficial attraction of this empiricist interpretation of concept formation is that "the concept does not appear as something foreign to sensuous reality, but forms a part of this reality; it is a selection from what is immediately contained in it." As he also points out, the generic concept produced by the act of abstraction is a "thing concept"; that is, the concept is conceived in terms of an abstract object. With such a notion there is a danger of reifying the concept, of giving it an independent reality as a particular thing along with the other particular things. While the concept of the 'type' does not *necessarily* imply a form of naive realism, as Mayr (1963) seems to suggest, it does seem true as a matter of fact that the 'typological' tradition has been prone to reify its abstract concepts.

It seems that if we are to deal with, or, more exactly, avoid, the 'taxonomic problem' without taking the Darwinist route, we need to consider the possibility of a different taxonomic practice, hence a different, nonempiricist, concept of kinds. I discuss this question in Chapter 5 where, following Driesch and Cassirer, I argue that biologists need to develop kind concepts which are more akin to those employed in the 'exact' physical sciences. We also need to consider the possibility of a different concept of essence. None of the realist philosophers mentioned here regards the characterisation of essences and the formulation of definitions as being achieved simply by a process of abstraction of what is 'common' from the empirically 'given'. Rather, they broadly follow Locke and make a distinction between nominal and real essence. The nominal essence is that set of properties, the manifestation of which is necessary for a thing to be identified as of a certain kind. The real essence is the nature of the thing, conceived as the 'hidden' structure which causally accounts for its manifest properties and *"provides the grounds. for the choice of criteria of individuation and identity"* (Harré and Madden, 1975, p. 17); that is, it grounds the properties which should be regarded as 'typical' of a thing of that kind. Thus 'real definitions' pertain to the real essences of things, but real essences are not given in experience; rather, they have to be constructed by the work of the theoretical imagination. I consider this further in Chapter 4.

From Nominalism to Darwinism

There is, however, a strategy which could be adopted in an attempt to save the notion of essential natures and typical forms in the face of empirical variation whilst retaining an empiricist standpoint; that is to multiply kinds or classes. This device, however, potentially contains the seeds of its own destruction, at least when employed by biologists, for the multiplication process could, in principle, continue until there are as many kinds or 'natures' as there are actual individuals. From here it is apparently a short step for a biologist, though a large step for a logician, to claim that there are in reality no classes, hence no 'natures' or natural kinds, merely individuals. Although the step may appear short, it involves a profound change in logical relations, for individuals and classes are instances of different logical types. Whereas classes are related to each other by inclusion, which is a transitive relation, individuals are related to classes by membership, an intransitive relation (see Buck and Hull, 1966; Hull, 1967, 1976). Nevertheless, either by means of a logical sleight of hand or as a result of an unwitting logical error, the step has often been taken and a nominalist position has been adopted on a number of occasions in the past, especially by biologists with evolutionary leanings; Buffon and Lamarck are famous examples (see Hull, 1967). Darwin (1859) also adopts an overtly nominalist stance at one point, though his advocacy of nominalism is, at least in certain respects, misleading in the context of his theory, as will become apparent below.

Darwin attempts to use the taxonomic problem of variation in an argument to the effect that species taxa are purely nominal kinds. Thus he claims that

no clear line of demarcation has as yet been drawn between species and sub-species. between sub-species and well marked varieties or between lesser varieties and individual differences. These differences blend into each other in an insensible series; and a series impresses the mind with the idea of an actual passage. . . . From these remarks it will be seen that I look at the term species, as one arbitrarily given for the sake of convenience to a set of individuals closely resembling each other, and that it does not essentially differ from the term variety. The term variety, again, in comparison with mere individual differences, is also applied arbitrarily. (pp. 107–108)

Darwin writes here (and elsewhere in the *Origin*) as though the facts speak for themselves and the world dictates that we must abandon the conceptual distinction between species, varieties and individuals. But the indeterminacy, indeed futility, of this kind of attempt to read an ontology directly, that is, *atheoretically,* from experience and the problems of material practice is evident since the non-Darwinists, for example Bateson (1894), use exactly the same kind of facts concerning empirical variation as evidence for the proliferation of 'typical' forms within boundaries taken to be real and specific.

However, when Darwin's argument takes a somewhat more theoretical turn, his claim that species taxa are merely nominal kinds is effectively replaced by a claim of an entirely different sort, so that his apparent nominalism acquires a different significance. The conclusion he draws from the theory of descent coupled with the experience of variation is, firstly, that individual organisms are, as a matter of fact, connected genetically or genealogically and, secondly, that "if our collections were nearly perfect, *the only possible arrangement* would be genealogical" (1859, p. 427, my emphasis). This view is partially prefigured in the quotation given above in the reference to "an actual passage." Thus, in the theory, species taxa are are not conceptualised as either nominal or natural kinds for they are not conceptualised as kinds or classes at all but as empirical objects, temporally extended populations of individual organisms related by descent from a "common ancestor." Logically speaking, we could regard this as a shift from the idea of species (and other) taxa as 'Aristotelian' classes defined in terms of intrinsic properties to a conception of taxa as equivalence classes. The individual organisms which are the members of such classes are equivalent with respect to the "common ancestor," and the relevant properties which provide the criteria for class membership are not intrinsic properties but relational properties – more exactly, *one* relational property: origin or descent. As Darwin puts it: "The natural system [of classification] is genealogical" (1859, p. 433). Thus while we still, logically speaking, have a system, it is *not* a system of *forms* since the intrinsic properties of individual organisms – which includes morphological properties – have no role or significance in such a system; they can, though they need not, be indefinitely variable. In this context, therefore, intrinsic properties are contingent with respect to identity while the relational property of descent is necessary, that is, essential.

Ernst Mayr's Radical Empiricism

Darwin's conception of species forms the basis of Mayr's famous 'biological definition' of the species category, which, in its reformulated version, asserts: "Species are groups of interbreeding natural populations that are reproductively isolated from other such groups" (Mayr, 1969, p. 26). Thus, Darwinism proposes a change in the meaning of the term 'species'. Whereas formerly it denoted a class of classes, now it denotes a class of empirical objects – a class of populations. Consequently, Mayr presents Darwinism as a conceptual revolution which involves a rejection of those views which supposedly inhibit scientific progress, namely, essentialism and typology and their replacement by "population thinking"; essentialism is "a major misconception that had to be eliminated before a sound theory of evolution could be proposed." However, in his employment of Darwinism as a stick with which to beat essentialism, Mayr espouses an extreme form of empiricism. As he explains:

The assumptions of population thinking are diametrically opposed those of the typologist. The populationist stresses the uniqueness of everything in the organic world. What is true for the human species – that no two individuals are alike – is equally true for all other species of animals and plants. All organisms and organic phenomena are composed of unique features and can be described collectively only in statistical terms. Individuals, or any kind of organic entities, form populations of which we can determine only the arithmetic mean and the statistics of variation. Averages are merely statistical abstractions; only the individuals of which the population is composed have reality. The ultimate conclusions of the population thinker and of the typologist are precisely the opposite. For the typologist, the type (*eidos*) is real and the variation an illusion, while for the populationist the type (average) is an abstraction and only the variation is real. No two ways of looking at nature could be more different. (1963, p. 5)

It would seem that Mayr regards this argument as definitive, for he has repeated it verbatim a number of times since its first publication (see, for example, Mayr, 1966, 1976) The replacement of essentialism by this kind of radical empiricism might lead us to think that biologists should revolutionize their practice and act after the manner of the sages of Lagado who, Swift (1726) informs us, were so philosophically enlightened that they had abandoned language entirely and conversed only by means of *things* which they carried around on their backs in large sacks. Alas, we do not live in revolutionary times and such a radical transformation of practice seems unlikely. Consequently, far from facilitating progress in biology, Mayr's views, if adopted, would bring that science to a dead stop.

Consider, first, the "unique features" of which organisms are supposedly composed and which the populationist wishes to describe in statistical terms. Features which are unique *cannot* be described at all, for description involves the use of general or kind terms: 'red', 'lanceolate', 'tibia', 'sepal' and so forth.

Consider, next, the "unique" individuals. If Mayr's claim regarding their uniqueness is taken literally, then such individuals will possess no common properties at all. If the world is irreducibly diverse, then a statistical description of the characteristics of a population would be impossible, for it would not have any characteristics. Alternatively, if it is argued that the "uniqueness" of individuals lies in a particular *combination* of properties, then a statistical description of the characteristics of the population with respect to some properties is, in principle, possible. But such a procedure involves an averaging operation on the set of *similar* properties – that is, the set of determinates under each determinable – possessed by the members of the population. Hence, the members of the population (or a part of it) are being treated as members of a kind, type or class defined in terms of the properties incorporated in the description. This is also the case if the "uniqueness" of the individual is conceived in terms of the actual 'value' of some particular determinate possessed by that individual. All description, including statistical description, presupposes that the "features" and individuals described can be identified as kinds or types of things. In fact, as described by Mayr, the activity of the "populationalist" seems to be that of constructing a particular kind of abstract model, a metriomorph, and the source and subject for a metriomorph is always a class; the model represents that class for certain purposes (Harré, 1970). The most notorious metriomorph is the model of the human family which consists of 2.63 (metriomorphic) children.

Consider, finally, Mayr's employment of terms. To make the point that individuals of the human species are unique, he has to employ a term which, from the traditional perspective, is a type, kind or class term, namely, 'human species'. It is things of this kind which are unique. The contradiction is self-evident.

Now, it is true that the formal contradiction is removed if we argue, as Hull argues (as will be shown) that the term 'human species' (or *Homo sapiens*) is not the name of a kind or class, that is, a general name, but is, rather, the proper name of a population of individual organisms which is itself, ontologically, an individual. Nevertheless, an epistemological problem remains, for such populations are not 'given' but have to be constructed in material practice by picking out or identifying the relevant individuals from the indefinitely large population of individuals of which the world is composed. We therefore need criteria which permit us to identify individual organisms as belonging to a species taxon. On the traditional view of taxa as kinds, the criterion of identity involves the intrinsic properties of the individuals and the idea that a kind can be characterised, at any rate partially, in terms of a 'cluster' of properties distinctive of, and common to, the kind. If Mayr's claim regarding the uniqueness of individuals is taken literally, then, as I noted previously these individuals need have no intrinsic properties in common. Consequently, on purely empirical grounds the criteria for identity can only be relational properties. Alternatively, and given that Mayr subscribes to Darwinian theory, even if the individuals do have common intrinsic properties then, on theoretical grounds, these must be deemed irrelevant. As I have indicated, in terms of Darwinian theory, the criterion for identity is the relational property of descent and intrinsic properties are, in this respect, contingent.

Thus, the question of criteria for identity is a central issue and in particular whether, or to what extent, the Darwinist criterion could actually be employed in material practice. I consider this matter in the following chapter where I argue that *some* conception of type or kind is epistemologically indispensable for picking out relevant individuals as well as for subsequently describing them. Further, I argue that this conception, even if it takes the (initial) form of a mere description of a 'stereotype' rather than a definition, must consist of more than a simple, unstructured list of properties.

I turn now to a consideration of the relations between "typological" and "population" concepts. I want to argue that for Mayr to *oppose* "typological thinking'' to ''population thinking'' in the way that he appears to do is mistaken and misleading. As I have noted, I accept that an empiricist ''typology'' cannot adequately represent the real structure of nature, though I also want to maintain that it does not *totally* misrepresent this structure. I am also prepared to accept that ''typological thinking'' has been replaced in the history of thought or, more exactly, in the thought of some scholars, by ''population thinking.'' However, in formal terms the one is not substitutable for the other for their goals are different and distinct (nor, incidentally, are the terms 'type' and 'average' synonymous as Mayr seems to claim). ''Typological thinking'' is a method – as it turns out, a rather inadequate one – employed in a preliminary attempt to construct a system of forms, and such a system (if there is one) will necessarily represent forms outside of time and space. By contrast, ''population thinking'' is a component of the Theory of Descent, a theory which purports to have causal, explanatory power and therefore necessarily represents organisms diachronically, situated in time and space. Whereas systematics deals with both actualities and possibilities, explanatory theory deals only with actualities. Representing a diversity as a system is not at all the same thing as explaining it, and formal derivation of possibilities is a different activity to theoretical explanation of actualities. Driesch (1908) makes essentially the same point in a discussion of systematics and history. As he observes: '' 'Existence', as understood in systematics, is independent of special space and of special time'' (p. 258) and systematics ''has to deal with the totality of the possible, not only of the actual diversities'' (p. 264).

Consider the 'rational system' of alkanes (the homologous series; see below). Given this system, we can derive, as a possibility, a substance with the general molecular formula $C_{35}H_{72}$. Consequently, this molecule necessarily ''exists'' as a member of the alkane series. However, whether this molecule can actually exist, and an explanation of how it comes to exist, if it does, requires reference to intrinsic and extrinsic material factors which are quite distinct from any formal considerations.

Mayr is thus mistaken to oppose the two 'forms of thought' as though they were alternatives between which we had to choose. Pateman (1987) has made essentially the same point in relation to the philosophy of linguistics where similar debates have occurred. In principle, the problem of systematics would remain if the theory of descent were to be abandoned tomorrow. Of course, there may be no system of forms, but the failure or, depending upon one's point of view, partial failure, of one method of representation does not, by itself, demonstrate that this is the case.

David Hull and Darwinian Realism

I turn now to a consideration of the "radical solution to the species problem." Although the position to be discussed was first outlined by Ghiselin (1974) and, indeed, has its basis in the Darwin–Mayr conception of species as empirical entities, its elaboration is due to David Hull (1976, 1978, 1981, 1984 and bibliography therein), whose subtle and illuminating arguments are required reading for anyone who wishes to understand the conceptual structure of Darwinian theory. As such, they provide an ideal basis for organising my discussion. Hull's arguments are quite complex and in order to make them comprehensible within a limited space I will first present a rather bald résumé of his position and then go on to consider some aspects of the arguments in more detail.

If evolution theory is to be a regarded as scientific theory, its statements concerning "species" must be statements about a class of real entities. One of Hull's central concerns is to reinstate a realist view of species taxa in opposition to the nominalist views derived by some from the taxonomic problem of variation. However, this is *not* a natural kind realism, for Hull argues that, ontologically, species taxa are not classes at all, hence not natural kinds, but individuals; for clarity I will refer to them in future as species-individuals. Hull argues, convincingly in my view, that species taxa are in fact conceptualised as individuals in the Darwinian Theory of Descent. He further argues that insofar as species (and other) taxa might appear, empirically, to be classes of the traditional kind – and in Hull's view many or most do not – then they are not theoretically significant classes. That is, there are no theories in biology in which the names of particular taxa feature in an uneliminable fashion; therefore, they are not natural kinds. This is a question I will return to in Chapter 4.

On this new view, the relation between an individual organism and a species taxon is not that of member and class but that of part and whole. It follows that 'higher' taxa are not more inclusive classes but sets (of different sizes) of species-individuals related by common descent. Since we are now dealing with individuals, identity is a function of origin. With respect to identity, therefore, the similarity of the individual organisms which comprise a species-individual is contingent. From this perspective, the names of taxa cannot be intensionally defined since they are not the names of classes or kinds but proper names which are rigid designators. Moreover, since species taxa are individuals, they cannot figure in scientific laws. Thus, on Hull's account there can be no *scientific* explanation of the properties, including morphological properties, which might appear to be specific to particular taxa; these are to be explained in terms of historical narrative (see Hull, 1975, 1984)

It is important to note that none of these claims implies, as Harré (1986) seems to suggest, that evolution theory has dispensed with natural kind concepts. Hull's rejection of natural kind concepts pertains only to species (and other) taxa; that is, particular species taxa are not themselves regarded as natural kinds which have individual organisms as their members. But particular species taxa are *instances* of a natural kind, the kind named "species." As Hull (1976, 1978 and, especially, 1984) has made clear, evolution theory is a scientific theory about "species," that is, the properties common to *all* species taxa. The theoretically significant kind (the natural kind) in evolution theory is, therefore, the species category – the class of all species taxa – which in the new view remains a class, though now a class of individuals rather than a class of classes as in the traditional view. It is, of course, arguable whether "species" is the name of a *single* class; fortunately, this problem need not concern us in the present discussion, which is exclusively concerned with taxa.

While I am convinced that Hull's analysis of the way in which taxa are conceptualised within Darwinian theory is substantially correct, I am not persuaded as to either the desirability or the necessity of Hull's suggestion that the new, Darwinist view in which species taxa are conceived as individuals and (implicitly) higher taxa as sets of species-individuals should be adopted as *the* conception of taxa within biology generally. This position, which is the "central subject" of all his later writings, is explicitly mooted in his 1976 paper, presented in a qualified form in his 1978 and expressed as a personal goal in his 1984: "I myself prefer to strive for a single univocal species concept applicable throughout biology. I think that species as historical entities functioning in the evolutionary process is the best candidate for such a univocal species concept" (p. 37).

I am not convinced by Hull's claim that such a "univocal species concept" would have significant advantages as compared with the traditional view. That, for example, it will "eliminate longstanding anomalies both within and about biology" (Hull, 1978). As I will argue, while the new view does have some advantages, its disadvantages are considerable, not least for a *science* of morphology. While Hull's position as regards species taxa is not nominalist but realist, this particular form of realism has consequences as regards the study of the forms of individual organisms that are virtually indistinguishable from those which would result from a nominalist position since in

Hull's ontology there are only two basic natural kinds, "species" and organisms.

As I noted earlier, I share Hull's (1976) view that the ontological status of entities is theory dependent and that classifications should be theoretically based or theoretically significant and, therefore, that theory always has the last word. However, I do not accept that this position is to be interpreted as meaning that theory has the *only* word, since, from my perspective, theoretical and material practice are conceived as related dialectically. Consequently, considerations of material practice must be given their due weight. Nor can I accept that theoretical considerations should predominate irrespective of the explanatory power or scope of the theory in question in relation to specific phenomena; in the context of this discussion, these are morphological phenomena.

Let me now elaborate some aspects of Hull's argument within which I wish to distinguish three main strands or components.

1. An empirical component consisting of claims concerning the taxonomic problem of empirical variation and the difficulty or impossibility of formulating definitions of the names of taxa. This has been outlined above.
2. A theoretical or theory-based component in which claims are made about the ontological status of taxa on the basis of how they are conceived to function in Darwinian theory.
3. A conceptual or metaphysical component consisting of relatively uncontroversial claims about the properties of individuals and classes together with claims about the nature of the part–whole relationship.

These components or strands are organised into an argument in the context of a particular position in the philosophy of science and consequently a certain view of the nature of scientific laws. This position is "a lineal descendant of logical empiricism" (Hull, 1976).

The core of Hull's argument (1976, 1978) concerns the way in which species taxa are conceptualised in the Theory of Descent. The argument is simple and convincing. Darwinian theory requires that species behave in certain ways and, whereas individuals can behave in the ways required, classes cannot. Thus, within the theory, particular species taxa are required to split since this is the mechanism whereby new species arise. Classes are not the kinds of entities which can split because properties do not have criteria of identity whereby they can be individuated and reidentified – a standard metaphysical thesis (see Carr, 1987). Individuals, on the other hand, are capable of splitting and do possess the requisite criteria of identity. Secondly, the theory requires that species taxa have a distinct origin in time and space and claims that, sometimes at least, they can become extinct, that is, cease to exist. Classes are not the kinds of things that can have distinct orgins defined in spatio-temporal terms because they are not spatio-temporal entities. Nor, for the same reason, can classes cease to exist. A class may have no members, but it nevertheless remains as a class. To use one of Hull's (1976) examples: if all gold atoms ceased to exist, the class of gold atoms would have no members, but a space would remain in the periodic table. If at a later date atoms with the appropriate structure came into being again, such atoms would, unequivocally, be members of the class 'gold'. An individual, on the other hand, does have a distinct, spatio-temporally defined origin and can cease to exist; once an individual has died, that numerical individual has disappeared forever. Finally, the theory claims that species diverge over time by means of the gradual and progressive accumulation of variations. This requires that species be spatio-temporally continuous entities. Individuals are such entities and classes are not.

Hull's argument concerning the ontological status of species taxa *as conceptualised within Darwinian theory,* is convincing. In terms of this theory, species taxa cannot be conceptualised as classes; they must be conceptualised as individuals – in Hull's (1978) gloss: "spatio-temporally localised cohesive and continuous entities (historical entities)." Now, Darwinian theory also asserts that (at least some) higher, more inclusive, taxa have a distinct spatio-temporal origin; thus, in terms of the theory, to regard a set of species taxa as included within a genus is to claim that the set has a common ancestor. It seems to follow, on theoretical grounds, that genera and (some) other higher taxa cannot be traditional classes either. Elsewhere, Hull (1965) argues on empirical grounds (the problem of formulating definitions) that the phylum Chordata cannot be conceived as a class. However, it is doubtful that higher taxa can be regarded as individuals since, as Hull (1976) observes, while spatio-temporal continuity – in Darwinian theory, integration in terms of descent – is a necessary condition for individuality, it is not sufficient; a degree of "unity" or "a certain cohesiveness" is also required. In the case of species taxa this requirement is satisfied by interbreeding, but it is not clear how it could be satisfied in the case of higher taxa which do not seem to possess any appropriate form of internal cohesiveness. As I have noted, in purely formal terms Darwinian theory proposes that higher taxa be conceptualised as equivalence classes. However, in terms of Hull's scheme, I would suggest that higher taxa must be regarded as sets (of different sizes) of species-individuals. The requirement for "cohesiveness" also raises problems about the exact relationship between species taxa as conceptualised in the Theory of Descent and the interbreeding populations of Mayr's 'biological definition' of species, because sometimes a population is coextensive with a species and sometimes it is not; here I refer readers to Hull's (1976) discussion since the issue is peripheral to my concerns.

Now, if species taxa are individuals, individual organisms must be conceived as *parts* of such species-individuals, which, therefore are *wholes*. Thus, on the new view, the relationship between individual organism and species taxon is not that of member and class but that of part and whole. Hull draws a number of conclusions from this thesis.

In the first place, if species taxa are individuals, then their names – *Fritillaria meleagris, Drosophila melanogaster* and so on – are not kind names but proper names. This conclusion seems to make sense of the difficulty, noted previously, which taxonomists have in formulating definitions of the names of species taxa, for it is widely accepted that proper names simply denote; having no sense, they cannot be defined. As Hull (1976) neatly puts it: "Species names cannot be defined in the traditional manner because they cannot be defined at all." In Hull's view, then, species names denote a "chunk of the genealogical nexus" and that is all.

Further, the identity of individuals is a function of their origin, not of any properties which their parts may possess. To use one of Hull's examples: identical twins possess similar parts but they remain numerically distinct individuals. Conversely, the parts of a numerical individual can change considerably during the lifetime of that individual – think of human development and aging or the complex life cycles of many organisms – but the individual retains its identity; it remains one and the same individual. Thus on the species-as-individuals view, the similarity of individual organisms is contingent with respect to the identity of the taxon of which they are parts. As it stands, this is a purely conceptual thesis which has no empirical consequences whatsoever with regard to the existence or nonexistence of similarity (of some kind) between individual organisms which are parts of a species taxon. All that is implied is that any similarity which might exist does not constitute a criterion of identity with respect to the taxon and in this sense is irrelevant; similarity is a "red herring" (Hull, 1978, 1981).

The conceptual claim concerning the empirical similarity of individual organisms in relation to the identity of a taxon is supplemented by a conceptual claim concerning similarity in relation to the identity of individual organisms conceived as parts of a whole:

Just as a heart, kidney and lungs are included in the same organism because they are part of the same ontogenetic whole, parents and their progeny are included in the same species because they are part of the same genealogical nexus, no matter how much they might differ phenotypically. The part/whole relation does not require similarity. (Hull, 1978; see also 1976)

Thus, in relation to identity, dissimilarity – that is, variation – ceases to be a taxonomic problem in the sense that that a taxon need not (though it might) consist of similar parts – that is, individual organisms – and it remains one and the same taxon despite any changes there might be in the parts. Likewise, individual organisms can be identified as parts of this taxon even if they are dissimilar to each other (though they might be similar).

In the next two chapters I present below a critical discussion of how this position relates to biological practice. Hull's general argument is that the "univocal species concept" is either consistent with current practice or makes sense of this practice. I consider, firstly, in Chapter 3, material practice, that is, taxonomic and experimental practice, and, secondly, in Chapter 4, theoretical practice; this will involve a consideration of the conception of scientific laws, explanatory theories and the status of 'natures'. Although my discussion is organised around Hull's demand for a "univocal species concept," the central question with which I am concerned is, as I have stated, the ontological status of taxa in general. Since, as I have noted, I believe that Hull's analysis of the conceptual structure of evolution theory is substantially correct, it follows that a critique of the conclusions which he draws from this analysis is, in effect, a critique of the view that the metaphysical structure of other areas of biology should be based on, or dictated by, the metaphysics of evolutionary biology. In other words, it is a critique of the view that evolution theory is *the* unifying theory in biology.

The Current Predicament

Let me attempt to summarise the main points of the argument so far. The dilemma that is presented by the Linnaean attempt to systematise forms empirically is that while the attempt cannot be regarded as successful, neither can it be regarded as a complete failure.

The nominalist and Darwinist traditions have emphasised the lack of success. Mayr (1969) exemplifies this position in his polemic against 'typologists': "Variation . . . is considered . . . as trivial and irrelevant. The constancy of taxa and the sharpness of the gaps separating them tend to be exaggerated" (p. 67). In similar, but more precise, vein, Hull (1976) argues that taxa cannot be conceived as classes: "No matter how hard they tried, taxonomists could rarely find sets of traits which divided living organisms into neat little packages" (p. 180).

Bateson (1894), on the other hand, emphasises the relative success of the enterprise. While the forms of living things are diverse

They may ... be separated into Specific Groups ... the members of each such group being nearly alike, while they are less like the members of any other Specific Group. ... The fact that in certain cases there are forms transitional between groups ... is a very important fact ... but it remains none the less true that ... between the immense majority of these [Groups] there are no transitional forms. This is true of the world as we see it now, and there is no good reason for thinking that it has ever been otherwise. (p. 2)

It is difficult to believe that the factual information available to Mayr and Hull and to Bateson is radically different. Rather, their positions represent different views of a genuine predicament.

The nominalist and Darwinist traditions within morphology have stressed the lack of success of the Linnaean enterprise via a concentration on the taxonomic problem of variation. Nominalists have claimed that classification is a matter of human convenience and convention. However, it is arguable that nominalism has always been a minority tendency within morphology and most morphologists have tended towards some form of realism as regards at least some taxa. Darwin himself seems to have held different and contradictory views simultaneously. However, it is clear that the Theory of Descent does not conceive taxa as natural kinds and positively claims that the order of nature is merely an historical order. Darwinian realism as explicated by Hull may be regarded as the culmination of the 'old dialectic' in that it presents a coherent and internally consistent view of the ontological status of taxa. However, I will argue that this form of realism cannot sustain some important forms of material practice in biology which presupposes that individual organisms under morphological descriptions are conceived as members of kinds. Moreover, an essentialism based on origins raises just as many problems as the form of essentialism that it is intended to replace; the individuation of origins is just as problematic as the individuation of species, not least because the origin of a species is another species. Lastly, I will argue that, in relation to theoretical practice, the mode of explanation of form which is permitted by this position either evades the problem of the genesis of such forms or, at best, permits only a partial and inadequate account. These facts in themselves provide some incentive for the development of a theory which can ground a notion of forms as kinds.

If we do not subscribe to nominalism and if we believe that the attempt to make use of the Darwinian form of realism in contexts outside evolution theory proper gives rise to real difficulties, we might wish to emphasise the measure of success of the Linnaean enterprise. Properties are not randomly distributed among individual organisms but show *some* tendency to cluster; individual organisms do resemble each other in differing degrees to *some* extent; there is *some* regularity and repetition of forms; it makes *some* sense at least *some* of the time to talk of 'normal' forms – most Crocuses do, as a matter of fact, exhibit a threefold rotational floral symmetry. It is implausible to suppose that the pattern of empirical order which has been revealed by some two hundred years of painstaking work is completely illusory. While biological forms do not fit neatly into conceptual boxes, there is *some* degree of fit, albeit an untidy one. Indeed, *some* degree of order of this type, the possibility of arranging organisms in "groups under groups" (Darwin, 1859), is presupposed by evolution theory since the theory attempts to explain it.

It would seem that debates at the level of empirical order have effectively reached an impasse. As Kitts and Kitts (1979) point out, "Whatever order is to be provided in terms of the manifest properties of objects and events has, in effect, already been provided. To impose new patterns of order scientists must point to previously unrecognised, which is to say underlying properties" (p. 617). Much the same position is implied by Bateson (1894). Kitts and Kitts are primarily concerned with evolution theory proper and the origin of reproductive isolation. They argue, contra Hull, that the origin of species must be understood as the origin of kinds and, in so doing, provide the outlines of a natural kind realism which is broadly compatible, although not identical, with that to be developed here.

It would seem, then, that if any progress is to be made with regard to the problem of kinds in relation to the problem of form, we must gamble on the possibility that the existence, in terms of one system of representation, of some degree of order at the level of appearances, even if it is only an order of resemblance, is a clue to the existence of a greater degree of order at a deeper level. Here it may be helpful to note the general argument advanced by Bhaskar (1978) that if some or all of the 'ordinary things' of the world are, metaphysically speaking, compounds, then it is understandable that they should share nothing in common empirically but resemblances. I suggest in Chapter 4 that adult organisms may be 'ordinary things' of this kind. Nevertheless, the genesis of these 'ordinary things' and the changes in them require explanation which, for realists, will be couched in terms of continuants and causes. Thus, explanation will make reference to 'extraordinary things' such as electrons, atoms, molecules and, I will suggest, morphogenetic fields, which share a common nature, therefore a common identity, that is, things which fall into a natural kind. As I argue in Chapter 4, natural kinds cannot be equated with empirical classes and "Scientifically significant generality does not lie on the face of the world, but in the hidden essences of things'' (Bhaskar, 1978, p. 227). The groupings which constitute classifications – humanly constructed but, nevertheless, intended to reflect a real, natural order – can only be made determinate in terms of the unobservable properties and structures of the beings in question.

And it may be some incentive to place a bet if we remember that, sometimes at least, this kind of gamble in the physical and chemical sciences paid off. We should remind ourselves that the science of crystallography succeeded in elaborating an elegant system despite the fact that many or most naturally occurring crystals are relatively irregular. We might also remember that in chemistry at least some of the kinds distinguished in a purely material practice, that is, pretheoretically, have proved to share important microstructural properties, that is natures (see Dupré, 1981). Samples were not originally included in the extension of 'gold' because of their nature (real essence) for the simple reason that this was not known. Likewise, the close similarity of the halogens was recognised in material practice long before theories concerning electron shells justified their adjacent positions in the periodic table. But we need also to remember the great length of time and the enormous expenditure of intellectual and practical labour which was required to elaborate the theoretical systematisation of the elements represented in the periodic table. The change in the explication of the concept of 'copper' from a ''red, malleable metal'' to a ''collection of atoms each characterised by the presence of single electron outside the filled 4^d shell'' required several hundred years (Harré and Madden, 1975). A *science* of form, supposing such to be possible, will not be created in six days.

3

The Ontological Status of Taxa: Material Practice

Taxonomic Practice

I begin with a consideration of the material practice of taxonomy since it is in this context that the practical difficulties posed by empirical variation take on significance. On the one hand, Hull argues that taxonomic practice is consistent with his position and as such provides empirical support for his view that species taxa should be regarded as individuals, and that the relation between an organism and a taxon is the part–whole relation. On the other hand, he argues that some aspects of taxonomic practice are clarified or rationalised by his position. I do not find these arguments compelling. In my view, taxonomic practice when considered along with experimental practice (which Hull does not discuss) does not provide strong support for the "univocal species concept" but rather speaks against it. I want to argue that irrespective of their current theoretical status, species, and possibly other, taxa are *practically significant kinds*. In other words, in some aspects of their practice, biologists treat species and other taxa as kinds and individual organisms as samples or examples of those kinds, and this in itself should provide sufficient incentive for developing a theoretically significant notion of kinds as natural kinds. Whether these kinds are to be regarded as traditional classes or in some other way is another matter, one which, I shall argue, can only be settled by appeal to a different, and as yet largely undeveloped, theory; a theory of morphogenesis.

Naming

I now present a relatively brief critique of some aspects of Hull's (1976, 1978) discussion of taxonomic practice. At the heart of taxonomic practice is a naming practice and Hull considers in outline how a species taxon acquires a name or, to put the matter another way, how a name acquires 'meaning'. According to him this involves the baptism of an individual organism (the holotype specimen) which is, in effect, chosen at random.

A taxonomist in the field sees a specimen of what he takes to be a new species. It may be the only specimen available . . . [he] could not possibly select a typical specimen . . . because he has not begun to study the full range of the species' variation. He selects a specimen, any specimen, and names it. Thereafter, if he turns out to have been the first to name the species of which this specimen is a part, that name will remain firmly attached to that species. A taxon has the name it has *in virtue of* the naming ceremony, not *in virtue of* any trait or traits it might have. If the way in which taxa are named sounds familiar, it should. It is the same way in which people are baptised. They are named in the same way because they are the same sort of thing – historical entities. (1978, p. 352).

A preliminary comment on this claim might be that a taxon cannot have the name it has *solely* in virtue of the naming ceremony since it is difficult to see how a taxonomist could suppose that he was dealing with a *new* species, and hence that a new and distinctive name was required, by any means other than the "traits" the baptised specimen might have. That is, the specimen possesses "traits" which can be distinguished from those possessed by specimens belonging to species already known. However, the main thrust of the argument is to suggest that taxonomic practice is consistent with an ontology of species-individuals and the concomitant view that the names of species taxa are proper names which function as "rigid designators," that is, terms which denote one and the same entity under different descriptions and in counterfactual situations; across "possible worlds," as the logicians have it. Now, this aspect of taxonomic practice is apparently consistent with Hull's position. On his view, what is being baptised is a part of a whole (an individual) which, in a sense, represents this whole. If parts are subsequently discovered which differ from the originally baptised part in traits (and hence are described differently) the original name is still applicable; the name designates "rigidly." Just so, the name Charles Robert Darwin was assigned to a particular infant – a temporal part of a whole, individual life – on 17 November 1809, and remained attached to the man who was buried in Westminster Abbey on 26 April 1882. The name stuck despite quite radical changes in the properties of temporal parts of the life subsequent to baptism, hence changes in the contingent criteria by means of which the individual might be identified or described: the beardless youth baffled by algebra; "the man who walks with Henslow"; the man who lost raven-haired Fanny Owen to another; the white-bearded student of earthworms; "the greatest Englishman since Newton" and so forth (see Desmond and Moore, 1991).

Thus, Hull's view of species taxa as individuals appears to rationalise the fact that, when taxonomists subsequently obtain greater knowledge of the range of variation in the taxon, the randomly chosen holotype individual may turn out to be 'abnormal'; the "type specimen" need not be "typical." On a conception of taxa as classes, this seems anomalous or paradoxical, whereas on Hull's view, an abnormal individual – a *Crocus* with a flower showing fourfold rotational symmetry – is as much a part of a whole as any other individual. Any position which attempts to retain a view of taxa as kinds will have to account for this fact.

All the preceding discussion notwithstanding, however, this aspect of taxonomic practice is at least partially consistent with the view that species taxa are natural kinds and their names are kind or general names, for both Putnam (1975) and Kripke (1980) have argued that kind names also acquire their 'meaning' by means of acts of ostension and baptism involving individual samples and that such names also function as 'rigid designators'. If these arguments are accepted, the presence of an act of baptism and rigid designation is inconclusive with respect to Hull's position.

Hull is, of course, aware of the possibility of this kind of counterargument and comments (in a footnote, 1978) that "Kripke's analysis [of general terms] is controversial." It is true that the views of Kripke and the closely related views of Putnam are controversial (see Harrison, 1979; Dupré, 1981; Harré, 1986). But arguably they are no more controversial than Hull's analysis of species names. Moreover, as I will argue below, some of the criticisms which have been advanced against the Kripke–Putnam thesis are of a similar kind to those which can be advanced against Hull's own views.

As Harrison (1979) has pointed out in a general discussion of naming, a theory of ostension and baptism only explains how a name gains currency. These acts *by themselves* do not suffice to fixes a meaning for a name in the sense of establishing the category in logical grammar of the name in question – that is, whether it is intended or interpreted or used as a proper name or a kind name. Compare, for instance, the baptismal act in the dockyard: "I name this ship 'The Brighton Belle,' " with a similar act in the boardroom of Ford: "Let's call this car the 'Fiesta Popular Plus,' " or in the laboratory: "I vote we call this element 'Praseodymium.' " As regards the matter at hand the question can be rephrased in terms of the exemplary aspect under which an individual organism (the holotype specimen) is baptised; is this individual conceived as a sample or example of a kind or class, in which case the name is a kind or general name; or, alternatively, is it conceived as a part of a whole in which case the name is a proper name.

As Harrison puts it:

ostensive definition *in itself* does not suffice to fix a meaning. It fixes a meaning for a word *w* only when taken in conjunction with the form, or structure, of the investigation which links one sample ostensively labelled as *w* to other samples so labelled. ... Ostensive definition cannot give meaning to a name until we know what we are to do with the name: until we know, in the later Wittgenstein's terms, what language game the name is to be fitted into. (1979, p. 161)

Therefore, a study of practical use should reveal how biologists conceive the names they employ, hence what ontological status, albeit provisional and defeasible, they give to the entities named. I argue here that taxonomic practice might be interpreted as giving some support, and subsequently that experimental practice gives stronger support, to the view that the names of taxa are conceived in practice as kind names and that to adopt any other position leads to major difficulties.

Extension

As is clear from Harrison's remarks, a central aspect of the practical use of a species name is its extension. Although this question is not addressed in great detail by Hull, his various remarks on identity suggest what, on his view, the basis of extension ought to be. On some variant of the traditional view that species taxa are kinds and individual organisms are members or examples of these kinds, extension should be based on some form of intrinsic similarity between the relevant individuals. On Hull's account, however, since individual organisms are parts of a whole which is an historical entity, extension should strictly be based on the theoretically significant "biological" relations between the individuals concerned. Although interbreeding is sometimes cited as a criterion for identity, it is not strictly relevant to the identification of individuals but rather serves to determine the categorial status of a taxon; for example, whether it is to be regarded as a species taxon (Hull, 1969, p. 244; see also van der Hammen, 1986). The relevant relation with respect to identity is that of 'descent', which, on Hull's view, is synonymous with spatio-temporal continuity between any given individual and a parental organism. Hull makes this clear by quoting approvingly from Griffiths: "[The] reference of an individual to a species is determined by its parentage, not by any morphological attribute" (Hull, 1978, p. 349). Immediately following this remark, he makes a comparison between organisms and chemical elements. In the case of chemical elements, such as gold, the origin of a particular atom of gold is irrelevant to its identity as gold, which is dependent upon its atomic structure. As regards the identity of chemical elements, therefore, origin is contingent and properties are necessary. Hull implies that the converse is the case with organisms. As he puts it, "In the typical case, to *be* a horse one must be *born* of horse" which, as I read it, implies that, as regards identity, a particular origin is necessary. This of course would be consistent with the evolutionary view that a taxon has a single, spatio-temporally localised origin. In this context, it appears that "to be a horse" primarily pertains to having a particular identity and not to having a particular set of empirical characteristics; as regards identity, properties are contingent. Thus, whatever Bellerophon may have thought, Pegasus was no horse, not even an abnormal (or magical) horse. For we are told that he was born not of horse, but from the blood of Medusa. That a particular origin is necessary is confirmed by Hull's remark on the same page that "if a species evolved which was identical to a species of extinct pterodactyl save origin, it would still be a new, distinct species" (see also Hull, 1976).

Now, it is a fact that biologists make use of relations of origin or 'descent', that is, spatio-temporal continuity, to substantiate claims that distinct, and often markedly different, entities are parts of the *same* whole – for example, 'stages' in the life cycle of one and the same individual. Biologists also employ relations of 'descent' to substantiate claims that two individuals belong to the *same* species taxon – for example males and females which may differ markedly in overt morphological characteristics. These facts were, of course, recognized by Darwin (see 1859, p. 407), and used by him as part of the argument that a 'natural' scheme of classification should be genealogical. Superficially, they appear to support Hull's position, but closer inspection shows that matters are not so straightforward.

In the first place, we might want to question whether, in both a theoretical and a practical context, a relation of descent can be regarded as synonymous with a relation of spatio-temporal continuity. Or, to put the matter somewhat differently, whether a relation of spatio-temporal continuity between two entities is necessary *and* sufficient for claiming that these entities belong to the same species.

Now, there are some entities which can justifiably be conceived as spatio-temporally continuous with individual organisms but which would not normally be regarded as related to them by descent; for example, excretory products. Thus, while spatio-temporal continuity is a necessary condition for the predication of a relation of descent, hence specific identity, it is not sufficient. What is required in addition is some form of similarity. At the very least, the putative descendant must be an entity of the same general kind as the putative parent: that is, an organism.

However, there are some organisms which might, in the practical context, be conceived initially as spatio-temporally continuous with other organisms even though they are not; certain kinds of parasitic organisms are obvious examples. It can be argued that the practical investigation of the true nature of the relations of such organisms involves a double activity, in that the determination of the relations of spatio-temporal continuity goes hand in hand with the recognition and separation of the life cycles of the entities concerned. That is, in practice, one entity, *Y*, is viewed as related by 'descent' to some other entity, *X*, not only because there is a relation of spatio-temporal continuity between the two but also because *Y* eventually gives rise to a third entity, *Z*, which is similar to, that is, of the same specific kind as, entity *X*. Another entity *P* is not regarded as being so related to *X* because it does not give rise to another entity of the kind *X*. This 'giving rise to' need not be immediate; several other entities can intervene, and in complex life cycles they do. What is required is the occurrence *at some time* of a form which can be regarded as a *repetition* of an earlier form. In practice, talk about 'descent' seems to involve talk about repetition or similarity, therefore putative kinds, as well as talk about spatio-temporal continuity Reference to kinds in a minimal sense is necessary, and in the practical context, the sense may be more than minimal.

The dialectic between considerations of descent and similarity can be seen in Darwin's (1859) discussions, for despite his insistence on the significance of relations of descent, he will not accept relations of this kind as the only grounds for predicating specific identity. Thus he claims that varieties belong to the same species as the "parent form" not only because they are descended from it, but also because they resemble it (p. 407). Moreover, he is sometimes prepared to override the criterion of descent, even if this leads to apparent inconsistency. In a possible-worlds scenario, he considers the case of a kangaroo born of a bear (p. 408). Is the offspring to be ranked as a bear? It would appear that he thinks not and therefore believes that origin is not a sufficient condition for taxon membership. But he evades the conceptual issue by claiming that the case is empirically "preposterous"; individuals related by descent will, as a matter of fact, closely resemble each other. In making this claim, he appears not to notice that it apparently contradicts his earlier claim concerning the primacy of descent in identifying the individuals that comprise "alternate generations" which do not resemble each other. The fact that Darwin intuitively regards the kangaroo–bear case as preposterous and the alternate generations case as not suggests that, in the practical context, he is only prepared to regard relations of spatio-temporal continuity as sufficient for taxon membership when dealing with individuals which, on other grounds, are regarded as belonging to kinds less inclusive than the kind 'organism'.

The argument presented here suggests that, in the practical context, there is more to descent than spatio-temporal continuity. However, a more direct argument can be presented against Hull's position in the context of taxonomic practice. We can accept that relations of 'descent' (in the qualified sense outlined) may serve to establish that two individual organisms are members or parts of the *same* species taxon. However, relations (of any kind) between any pairs of individuals cannot serve as the means for establishing the identity of a particular individual – in the sense of establishing to *which* (named) taxon that individual belongs – without begging the question. For to say that a particular individual is (can be known as being) a horse because it is descended from, is similar to, or even interbreeds with another horse is to presuppose horse and therefore to set up a potentially infinite regress (see Putnam, 1975). The formal role of the holotype specimen in taxonomic practice must be that of bringing this regress to an end. Thus, in the last analysis, the basis of extension must be the relation between *the holotype individual* and other individuals.

It is evident that in this, the pertinent situation, relations of 'descent' cannot serve as a practical basis for the extension of the specific name and therefore the delimitation of the taxon, for in the overwhelming majority of cases where the holotype individual is a specimen from the wild it is impossible to ascertain directly the relevant relations between the holotype individual and *any* of its ancestors, descendants, collateral relatives and their descendants which together comprise the extension of the name. At best, the name might be extended to a *few* of the immediate relatives of the holotype individual in situations of 'domestication' where a breeding programme had been established and a clone or strain had been produced from, or involving, this individual, but such situations would be very rare. Thus, an interpretation of Hull's position where taxonomic practice is to be organised on the basis of the theoretically significant relations between organisms seems to imply that taxonomy is impossible.

If relations of filiation or descent per se cannot serve as the practical basis for the extension of the name of a species taxon from the holotype individual to other individuals, we are left with some form of similarity as the basis and, of course, observable similarity or resemblance *is* used by taxonomists as the basis for extension. In the case of higher taxa, it is the only basis in principle which could be used. But the possibility of using resemblance in this way presupposes the existence of *regularity* in the life cycles of the individuals which comprise the taxon, that is, the repetition of similar, indeed typical, forms. Thus, although the holotype individual may not, indeed cannot, be "typical" in the sense of somehow physically representing the whole taxon, it must be "typical" in the sense of being a typical part of a life cycle so that comparisons, either direct or via descriptions, can be made between it and parts of other life cycles. If it is not typical (or typical enough) in this sense, then it cannot play the role which is required of it. In this event, the museum or *hortus siccus* amounts to little more than a cabinet of curiosities.

It would appear, therefore, that whatever their theoretical views, in taxonomic *practice* biologists have no option but to treat the names of species taxa *as if* they were kind names. The present line of argument, when taken in conjunction with the preceding one on the requirement for the repetition of qualitatively similar forms in at least some predications of 'descent', is, therefore, at least consistent with the position that the names of species (and other) taxa are practically significant kind names.

We can, of course, retain the Hull account in which the names of species taxa are proper names, but at a cost. To do this we have to argue that the resemblance or approximate similarity between the holotype and other individuals is being used as *evidence* for the theoretically more fundamental 'biological' relation of descent, or more generally of a common "genealogical nexus" (see Hull, 1965). Indeed it might be argued that, in this respect, the practice of biological classification would not differ radically from the classification of, say, the chemical elements since in this situation also the theoretically relevant similarity of atomic structure is not directly accessible but has to be inferred on the basis of evidence gained in the practical context. However, there is a fundamental difference between the two cases, for on Hull's view, as noted, similarity of properties is contingent with respect to identity. If we adopt this position, therefore, we have to accept that the extension of the name, hence the identification of new individuals, is indeterminate. This also means, of course, that the taxonomists claim that a particular organism is a member (or part) of a *new* species – the starting point of Hull's discussion – is also indeterminate. Identification can only become determinate if properties, or some of them, are conceived as typical for organisms of a kind. Consequently, we might be better advised to seek a different perspective in which taxa can be theoretically grounded as kinds.

"Homologous Traits"

In a response to his critics, Hull (1981) makes explicit his claim that, with respect to "the individuation of entities and properties," the criterion of identity is not similarity but "some relation which requires descent." He explains,

Traditional evolutionary biologists consistently insist on the primacy of descent. Taxa must be monophyletic, traits homologous . . . two traits may be as similar as one might wish, but they cannot count as the "same" trait unless they share a common evolutionary origin. . . . The relevant "identity" is identity through descent. (p. 144)

I presume that what Hull has in mind is the so-called Darwinian concept of homology. This is defined by de Beer (1932, p. 478) as follows: "The sole condition which organs must fulfil to be homologous is to be descended from one and the same representative in a common ancestor." The question thus arises as to what is the relation between this "Darwinian concept of homology" and the pre-Darwinian notion of homology, understood as morphological correspondence, which I discussed in Chapter 2.

The matter has been considered in detail by Woodger (1945), who points out that what de Beer's claim amounts to is that a part X in one form and a part Y in another form are homologous if, and only if, there is a third form which is ancestral to the first two and which contains a part Z which is in morphological correspondence with both X and Y. However, Woodger explains that when two parts X and Y are homologous in de Beer's sense, they are also in morphological correspondence with each other. Since the correspondence relation is both transitive and symmetrical, it follows that if X and Y are both in correspondence with an ancestral part Z, then X is in correspondence with Y. Thus both notions of homology involve the idea of morphological correspondence. The question therefore becomes, What is gained in the context of material practice by adding the 'evolutionary postulate' to the traditional definition of homology? The short answer is, nothing. Descent from a "common ancestor" is assumed, not observed. Consequently, de Beer's "sole condition" can rarely, if ever, be satisfied in practice. Nevertheless, as Woodger observes, morphologists do not hesitate to claim relations of homology (correspondence) between parts, for example, the endostyle and the thyroid gland, where there is no empirical information whatsoever concerning relations of descent between the individuals which manifest these structures. Contra Hull, in the practical context, the relevant identity is not identity through descent.

In a discussion of the legitimacy of "phylogenetic classification," Huxley (1942, p. 395) claims that it is "impossible" to maintain that structures such as the "pentadactyle limb or the crustacean appendage or the chordate notochord" have come into being on more than one occasion and so the "phylogenetic concept of homology" remains valid. Given our meagre theoretical understanding of the genesis of such structures, it is difficult to see how dogmatic empirical claims of this kind could be justified. However, this is not the main point for the issue is not empirical. Responding to Huxley's claim, Woodger emphasises that "there is no such thing as the pentadactyle limb, or the crustacean appendage, or the chordate notochord *apart from* the determining Bauplans of the Vertebrata, the Crustacea and the Chordata respectively" (p. 112). It is on the basis of these Bauplans that the morphological correspondences between the parts concerned are established, hence the claims of "sameness" made. In other words, the operational criterion for identity is a special kind of similarity, and anything which fulfils this criterion is an entity of this kind irrespective of any relations of descent. It is in terms of the relations between these Bauplans that classifications are constructed, and it is in terms of these classificatory procedures that attempts are made to *reconstruct* phylogenies since, in any situation of interest, "common ancestors" are not given as such. Rather, a particular form has to be judged as an ancestral form using exactly the same criteria of Bauplan and morphological correspondence as are employed in judging two contemporary forms as related.

As Woodger notes (p. 108), the advent of the theory of evolution altered neither the *practical definition* of homology as morphological correspondence nor the *practical procedures* of taxonomists. The practical business of individuating "homologous traits" continues to be based, as it always was based, on similarity, not on descent.

In sum: the nature of taxonomic practice suggests that whereas Hull's thesis has some virtues it also has defects. Overall, I believe that the analysis of this practice does not offer strong support for Hull's "univocal species concept" nor for the implicit claim that the traditional view of higher taxa as kinds should be discarded.

Experimental Practice

I now wish to examine experimental practice, where I believe the conclusions can be rather more definite as regards the ontological status of the entities identified and named in taxonomic practice. Here I develop an argument to the effect that a view of experimental practice as a rational, intelligible activity whose goal is the acquisition of knowledge of the causal mechanisms operative in nature presupposes that we conceive the individuals which are the objects of that practice as samples or examples of natural kinds.

Consider a case from classical experimental embryology. We observe in, say *Triturus taeniatus,* that the formation of the primary embryonic axis and the central nervous system involves a regular or uniform association (consequent upon gastrulation) between the mesoderm and the dorsal ectoderm. We hypothesise that there is a nomic rather than an accidental connection here; that is, we suppose the mesoderm causes a transformation of the ectoderm, which transformation is therefore an effect. In the language of conditions (von Wright, 1971, 1974), we want to assert that the presence of the mesoderm is a sufficient and a necessary condition for the effect. But in order to assert this we have to be able to assert the corresponding counterfactual conditional statement; that is, on any occasion in which the mesoderm – the putative cause – was in fact present, if it had been absent, the putative effect – the transformation of the ectoderm – would not have occurred. Now in principle, there is no way of verifying the truth of counterfactual conditional statements (von Wright, 1974). Nevertheless, by the experimental manipulation of systems, the investigator can come "very close" to such verification (von Wright, 1971). Such manipulation involves putting the system "into motion" by producing the putative cause when it is absent and permitting it to remain absent (when it is) or removing it when it is present. This is the procedure followed in classical experimental embryology. For example, the dorsal lip (presumptive mesoderm) from an early gastrula is surgically transplanted to the ventral surface of another gastrula. We observe that the embryo which has received the transplant produces a secondary axis on the ventral side. We repeat the experiment on a number of individuals and obtain similar results. On the basis of these experiments we conclude that the transplanted mesoderm is causally sufficient for the formation of a secondary axis; the system is closed to other causes. We then supplement our experiment with an observation on control individuals which have not received a transplant, where we note that a secondary axis is not produced. On the basis of this observation we conclude that the transplanted mesoderm is causally necessary for the formation of a secondary axis. In this kind of experimental manipulation we have come as close as we can to the verification of the counterfactual conditional, which, in the context of this experiment, says that had the secondary mesoderm been present, when in fact it was absent, a secondary axis would have been produced. In practice, of course, this observation of a control is supplemented by further experiments involving surgical deletion of the mesoderm from individuals where it is normally present.

Experimental practice of the kind discussed involves doing something to some individuals and noting that something follows and refraining from doing something to other individuals and noticing that the same thing does not follow. We regard these different experiences as complementary. On this basis we reason that if we had done something to that set of individuals where in fact we did nothing, our experience then would have been the same as the experience we actually had when in fact we did that same something to a different set of individuals. But this reasoning involves a presupposition, namely that what is actual in one set of individuals is possible in another set of individuals, and thus that the one set is substitutable for the other set in all relevant respects. To suppose this is to suppose that both sets of individuals are members of the same natural kind; they have the same nature. In von Wright's (1971) terminology, the "states of affairs" which are the subjects of nomic connection and experimental manipulation must be "generic"; that is, they must obtain repeatedly. Since it is a convention (based on experience) of experimental practice in biology that experimental and control individuals should be members of the same species taxon, it would appear that this practice presupposes that species taxa are natural kinds.

If we examine the further point or goal of experimental practice we see that the same presupposition is involved. From a realist perspective, the point is not to compile a record of empirical regularities, many or most of which are dependent upon our interventions in nature (Bhaskar, 1978) but rather to gain knowledge of a world outside the experimental situation and independent of us; that is, to gain causal explanatory knowledge of the process of morphogenesis outside the laboratory. To this end, if we are realists, we suppose that the knowledge of the nomic nature of connections which we have acquired by experimental manipulation provides evidence for the existence of a causal mechanism whose nature is presently unknown but which is responsible for the necessity of the causal connection we have discovered in our experimental individuals. We further suppose that this mechanism exists in all individuals of the same species because we hope that, eventually, this will provide the means whereby we can explain the production of the primary embryonic axis during the normal (and abnormal) development of all individuals of that species. The projection of the existential claim from experimental individuals to all individuals of the same species once again presupposes that a species taxon is a natural kind. On a quasi-nominalist view of the type which Hull proposes, such projection is, of course, irrational.

I concede that it could be argued that in the particular case I have used as an example we may not be concerned with species-specific properties but rather with those which are characteristic of a higher taxon; properties of Urodeles, say, or perhaps Amphibians, or even Vertebrates. But this concession does not affect the general thrust of my argument, since, on Hull's evolutionary view, higher taxa are no more natural kinds than are species taxa. I conclude that experimental practice, together with the reasoning based on such practice, presupposes that individual organisms are members of natural kinds and in this practice individuals are treated as samples or examples of a kind, not as parts of a whole. If we take into account the conventions of this practice, these kinds are species taxa. An ontological thesis of the type proposed by Hull renders experimental activity irrational, unintelligible and pointless.

Meaning in Practice and Theory; Descriptivism and Essentialism

The thrust of my argument is to suggest that species and, possibly, other taxa are practically significant kinds in biology and that this poses difficulties for a position which wishes to make metaphysical generalisations on the basis of a theory in which taxa are not conceptualised as kinds. However, I suspect that Hull would not be impressed by an argument of this kind for, as I read him, he gives absolute primacy to theoretical considerations. Thus, he notes (Hull, 1975) that "one of the most fundamental tenets of contemporary philosophy of science is that the basic units in any discipline are defined in the context of the theories which govern ... that discipline." This reading is supported when one considers his critical remarks (Hull, 1975) on operationism, that is the view that the meaning of a concept or term is synonymous with, or is defined in terms of, the practical operations performed to apply it. He might well characterise the position I have outlined as a form of operationism and reject it on these grounds.

Now, while it is true that contemporary philosophy of science has correctly emphasised the importance of theory, not least as part of a critique of the traditional, positivist view, it is also true that in recent years a number of philosophers of science of a realist persuasion (see Bhaskar, 1978; Hacking, 1983; Harré, 1986) have argued the importance of practice. Scientists cannot be regarded as merely contemplative intelligences whose sole activity is the construction of discourses. They also do things; they are embodied and act in and upon the world. From this perspective, it is arguable that Hull's apparent opposition of theoretically and practically grounded 'meanings' of names is too rigid and and his belief in the primacy of theory too absolute.

Harré (1986) has provided an illuminating discussion of the use of natural kind terms in the physical sciences which can be used to elucidate the problems with which we are concerned. He points out that theory and practice must be conceived as dialectically interrelated, as complementary rather than opposed. Thus chemical kind names are controlled by two sets of rules, one derived from the practical techniques of experimental chemistry and the other from chemical theory, especially theory about atomic structure. Like Harrison, Harré argues that we have to think of scientific activity as something like a Wittgensteinian 'language game'. Thus, although, ultimately, it is theory which is decisive as regards the 'meaning' of denoting terms, since practical distinctions must be theoretically grounded, matters of practice per se are not ignored. Harré explains:

human decisions to classify natural beings are fixed by fiat in contexts of practice. However, it is a prime aim of natural science to try to find out whether the constitution or nature of those beings, as disclosed by work in a theoretical context, justifies making the divisions into practical kinds where practice has drawn them. 'Natural kind' is a concept which can be explicated only within the double framework of practice and theory. Every natural kind is located in both contexts, and cannot be understood by reference to either one alone. (1986, p. 99)

Harré notes that recent discussions of natural kind terms and essences has taken the form of a polar opposition between "descriptivists" and "essentialists." The former hold that the extension of a kind term is determined by the prescriptions that a being must satisfy in order to fall under it, whilst the latter claim that the extension is determined by the natures of the beings that fall under it. Thus in the case of water, to take a much used example, descriptivists would hold that for anything to be water it must be colourless, tasteless, liquid at ordinary temperatures and so on. Essentialists, such as Kripke (1980) and Putnam (1975), on the other hand, would hold that sameness of kind is sameness of real essence. In terms of current scientific theory, 'water' is the name of a substance whose real essence (chemical composition) is H_2O; thus anything which is not H_2O is not part of the extension of 'water'. The term 'water', therefore, is not associated with any determinate set of descriptions; it is a rigid designator. In the context of our biological discussions, Linnaean taxonomy, with its requirement for intensional definitions of the names of taxa in terms of empirical properties, would appear to be a form of descriptivism. Hull's position, on the other hand, is a form of essentialism since he appears to claim, on theoretical grounds, that the essence of a taxon is its origin; sameness of species is sameness of origin and a species name is not associated with any determinate set of properties which can be regarded as necessary for the entity which bears the name. Thus Hull's position is comparable in certain respects to that of Kripke and Putnam. It is true of course, that for Hull species taxa are individuals rather than kinds; consequently their names are proper names rather than kind names. In this respect his position is similar to that of Kripke, who proposes, as regards individuals, a necessity of origin coupled with a contingency of development. However, the essentialist thesis of Kripke and Putnam is that kind names are *like* proper names in that *both* are rigid designators.

Now Harré argues that this opposition between descriptivists, with their emphasis on the practical context, and essentialists, with their emphasis on the theoretical context, is misleading; the two contexts are not opposed but complementary. Thus while Kripke and Putnam are correct in their insistence that science is concerned with real essences, their claim that the meaning of natural kind names is to be understood in terms of the rigid designation of real essences downplays the role of material practice in determining their use. Harré develops this argument by a consideration of Putnam's (1975) celebrated discussion of the liquid which exists in a possible world, "Twin Earth." I shall elaborate on Harré's argument and compare Putnam's views with those of Hull.

Putnam posits the existence on "Twin Earth" of a liquid called "water," which does not have the chemical composition H_2O but "is a different liquid whose chemical formula is very long and complicated. I shall abbreviate this chemical formula simply as XYZ. I shall suppose that XYZ is indistinguishable from water at normal temperatures and pressures. In particular, it tastes like water and it quenches thirst like water" (p. 223). Is the liquid really water? Putnam claims that theoretical considerations must take precedence over practical considerations. The liquid in question does not have the correct nature or real essence, kind names designate rigidly, therefore it is not water. Hull's argument, to which I referred previously, concerning the newly evolved species of pterodactyl identical in all apparent respects to an extinct species, is comparable to Putnam's argument. For Hull, too, theoretical considerations play the dominant role. The new species has a different origin to the extinct species, that is, a different essence, and species names designate rigidly, therefore it cannot be the same species as the extinct species, appearances notwithstanding.

Now Putnam's formulation of his argument is, perhaps, somewhat ambiguous. If he is claiming that the "Twin Earth" liquid is completely indistinguishable from water in the practical context, then it is impossible to see on what grounds we could ever assert that it had the composition XYZ rather than H_2O (see Dupré, 1981). Moreover, the claim that two substances could have identical phenomenal properties while differing in real essence undermines the explanatory role of scientific theory. Part of the point of claiming that water has a particular molecular composition is to explain why water has the phenomenal properties that it does. The statement "Water is H_2O" expresses a (presumed) linkage between the practical and theoretical contexts. In a similar fashion, Hull's assertions concerning the apparently identical pterodactyls which have different origins might be thought to undermine the explanatory claims of evolution theory since, in Darwin's words, "Propinquity of descent [is] the only known cause of the similarity of organic beings" (1859, p. 399).

On the other hand, to return to Putnam, the claim might be that the liquid on ''Twin Earth'' is practically indistinguishable from water in most but not in all respects, so that we could, in principle, determine that it was not H_2O. I suspect that this is what Putnam intends. But, as Harré notes, our response to this finding would depend upon its exact nature. If the liquid turned out to be a compound of curious and esoteric elements, we might well follow Putnam and decide that it was not water. Indeed we might go further and begin to question some of our assumptions about the relation between molecular composition and phenomenal properties. But if the liquid in question turned out to be deuterium oxide or tritium oxide, then we could quite happily allow practical considerations to dominate and refer to the ''Twin Earth'' stuff as 'water' on the grounds that it was chemically indistinguishable from water. And this practically based identification in terms of chemical behaviour could be theoretically justified since the different forms of water can be regarded as variants of a single natural kind (see subsequent discussion).

The response of biologists to Hull's pterodactyl or some similar case concerning a life form on another planet would, I suspect, be somewhat similar to that of chemists confronted by the waterlike liquid on ''Twin Earth.'' They would be exercised by the apparent identity of the new form and the earlier (or Earth) form and would feel obliged to make some sense of this. One response might be to attempt to demonstrate that the apparent identity was superficial or illusory and that there were, in fact, subtle but real differences between the properties or behaviour of the two forms so that, on practical grounds, they would not be entitled to the same name. If the attempt to achieve a practical distinction failed, there might be a more radical response which would involve calling into question evolution theory and the principle of monophyly. What I think biologists would *not* do is to ignore the problems in the practical context and content themselves with conferring a new name on the creature.

As Harré observes, while practical considerations are important, it is ultimately the theoretical context which controls the way kind terms are used. Distinctions made in the practical context and using descriptive criteria provide the initial impetus for the development of theories. But the legitimacy of practical distinctions must be grounded in terms of a theory about the real nature or essence of the things named which explains the production of the behavioural or property differences which constitute the basis for practical distinctions. Hence the dialectical relation between theory and practice. The essentialist error is to take the theoretical context as absolute and to downplay the role of practice, whereas that of the descriptivists is the converse.

The case of isotopes illustrates this point and Harré (1986) uses as an example the treatment of the isotopes of chlorine. From a descriptivist perspective, where distinctions between chemical elements are rooted in a practical context, the two isotopes, chlorine 35 and chlorine 37, are given the same name because they are indistinguishable in terms of chemical behaviour, that is, in terms of descriptivist criteria. From a rigid theoreticist perspective, however, they should be regarded as different natural kinds since they differ in atomic structure and therefore possess no common nature (real essence) in the strict sense. Regarding them as simply two variants of the same natural kind despite this fact can be theoretically justified since theory claims that chemical (as opposed to physical) behaviour – the basis for the original identification – is determined by electronic rather than by nuclear structure. The identity of the two isotopes formulated in the practical context of chemical behavior is therefore theoretically grounded in terms of an identity of structure which is *relevant* to *that* particular practical context; differences of structure (that is, nuclear structure) which are irrelevant to *that* practical context are effectively ignored as regards identity and natural kind status. Thus the two isotopes are both taken to be chlorine not because of an identity of chemical behaviour per se, but because this identity can be grounded in a theoretically relevant identity. The fact that theory is ultimately decisive is revealed by the case of the rare earths, for although these are difficult to distinguish practically in terms of their chemical behaviour, they are not regarded as variants of the same natural kind but as distinct elements because they are believed, theoretically, to differ in electronic structure, though this difference is supposed to exist only in an inner shell which has a minimal effect on chemical behaviour. As Harré argues, examples like this illustrate how the practical and theoretical contexts interact. Theoretical considerations such as electronic structure are relevant to the extent that they explain differences in behaviour in the practical context. Conversely, similarities and differences of behaviour count as significant in relation to identity and the distinction of kinds only if they can be theoretically grounded. Thus, although theory always has the last word, practical considerations are by no means irrelevant in the determination of kinds. We might also note that these examples from chemistry, arguably a relatively simple science as compared with biology, also teach that 'natures' and 'natural kinds' must not be conceived in a simplistic, all or nothing fashion.

Many of the chemical elements were distinguished in a practical context long before anything was known about atomic structures in terms of which the practical distinctions can be justified. Thus, as Harré notes, although the term 'gold' does denote a metal with a particular atomic structure (real essence), this cannot be its only denotation; it must also pick out a metal which has specific properties such that it can be manipulated in particular ways in particular forms of material practice, refining and jewellery making, for example. Likewise we might argue that *Solanum esculentum* and *Solanum nigrum* cannot be, as they are for Hull, *just* the names of two distinct chunks of the genealogical nexus. They must also be, to take a very practical context, the names of, respectively, a set of plants whose fruit is very good in moussaka (aubergine), and a set of plants whose fruits if employed in a similar fashion would give rise to a severe and possibly lethal stomach ache (black nightshade). That is, they must be the names of two taxa which can be distinguished in terms of descriptive criteria. I argued earlier in this chapter that it is principally descriptive criteria rather than theoretically based criteria (origin and descent) which govern the use of the names of taxa in biology, for it is on this basis that identifications are actually achieved and taxa constructed in the material practice of taxonomy. A central question, therefore, concerns the status of those properties which are referred to in descriptions and definitions of taxa. Are these merely contingent or accidental generalisations about the properties of individual organisms, as Hull would argue, or are they, or some of them, to be understood, as *typical* properties of kinds of organisms, to be explained as realisations of dispositions which are grounded in the natures of these kinds? In the present chapter I will examine this question by considering Hull's remarks on descriptions and definitions. In Chapter 4 I will pursue the question via a consideration of his view on the nature of biological laws.

Descriptions and Definitions

In his discussion of taxonomic practice, Hull (1978) raises the question of the role and status of the "descriptions," "diagnoses" and "lists of traits" which taxonomists attach to the names they introduce. He asserts that such "lists" are *descriptions*, they are not *definitions* – that is, sets of necessary and sufficient properties for membership of a class. I take it that this point is put forward as part of the argument that species taxa are historical entities or individuals, for whereas the names of classes or kinds are, on the traditional view, intensionally defined, individuals are described (see Hull, 1976) and descriptions "change through time as the entities which they describe change" (Hull, 1978). Thus, the name is not synonymous with the description. The example I gave above of the different descriptions applicable to that historical entity, Charles Darwin, at different times of his life illustrates the point being made. Likewise, the characterisation of, say, *Crocus cartwrightianus*, in terms of the threefold rotational symmetry of the flower is a mere description of a part of a particular whole. Since, on Hull's view, it is not necessary for an individual organism to possess this particular form of symmetry in order to be a part of this particular whole, the appearance in my garden of a plant of this species with a flower showing fourfold rotational symmetry creates no *taxonomic* problems. There is, of course, the problem of explanation.

Two questions can be raised with respect to this argument. Firstly, on Hull's view the description of a taxon must, initially at any rate, be the description of an individual organism, the holotype specimen, and this description must be an epistemologically useful description in that it enables us to pick out new individual organisms which are parts (or members) of the taxon.

Now, the concept of an organism per se is a very abstract concept, and I find it hard to conceive how a description of an organism which is epistemologically useful can be given which does not describe that organism more concretely as a *kind* of organism. Taxonomic descriptions do describe organisms in terms of kinds. They refer, for example, to the length of a *stem*, the shape of a *leaf*, the symmetry of a *flower*. In other words, an organism is described in terms of the properties of *specific* parts and the nature of the parts referred to in the description implies that the organism is an organism of a particular kind, in this case a flowering plant, because these parts are characteristic, or better perhaps, "stereotypical" (Putnam, 1975) of that kind. If the reference to specific parts is omitted, the description becomes epistemologically useless, not to say unintelligible.

Now, part of Hull's argument that species (and other) taxa are not classes, hence not natural kinds is, as noted earlier, that taxonomists cannot produce lists of empirical properties which can serve as the intensional definition of the name of a taxon; taxa do not have essences in this sense Hull claims. Rather, the names of taxa are defined as cluster concepts; that is, in terms of "sets of statistically covarying properties arranged in indefinitely long disjunctive definitions" (Hull, 1965, p. 323).

However, as Harrison (1979) argues, sortal or kind or class concepts cannot be equated with simple, unstructured lists or sets of properties, irrespective of whether such lists are conceived as descriptions or definitions and even if, in a Lockean manner, we suppose that such properties have a 'union in nature', that is, are conjoined in experience. Such a list is epistemologically inadequate in that we cannot use it to apply the name in question to some new entity unless we also know of what sort of thing the listed properties are supposed to be the characteristic properties; unless, as Harrison puts it, we have "an account of the *sort* of 'union in nature' which the properties ... must exhibit if they are to be taken as jointly characterising a sample of [say] gold" (p. 40). In other words, in order to identify a new sample as a sample of gold, we need to know not simply the characteristic properties of gold, yellowness, malleability, metallic lustre and so forth, but also that gold is the name of a particular *kind* of stuff or substance rather than, say, the name of a set of environmental circumstances which might also have a 'union in nature' (see Harrison, 1979, p. 39). A simple list of properties is inadequate to determine the reference of gold, hence the extension of the name, and we can identify new samples only by making use of a sortal concept of higher generality, in this case, kind of substance or kind of metal.

Now, on something like the traditional view, the list of intrinsic properties associated with the name *Galanthus plicatus* are the properties of a particular kind of snowdrop (*Galanthus*), a particular *sort* of 'union in nature'. Identifying a new individual organism as a sample of *Galanthus plicatus,* that is, as a sample of the same kind as the holotype specimen therefore involves identifying the entity as a sample of *Galanthus*. Likewise, knowing that *Galanthus plicatus* is the name of a kind of snowdrop, and knowing what individuals of this kind look like, is of *some* assistance in picking out individuals of *Galanthus nivalis* just as knowing that gold is the name of a kind of metal and knowing the properties of this kind is of *some* help in picking out a sample of silver. In both case the field of possibilities is reduced. We can further note that this 'union in nature' cannot be adequately characterised in terms of a mere set of properties, an unstructured collection, anymore than the Krebs cycle can be adequately characterised as a set or collection of chemical reactions. It has to be characterised in terms of a form or pattern of organisation; a whole with its parts and a certain 'arrangement'.

Thus, with respect to the material practice of referring or 'picking out', the binomial system of nomenclature seems to make epistemological sense if taxonomy is interpreted from a perspective in which the names of taxa are the names of kinds. On Hull's view, the names of taxa are not defined in terms of intrinsic properties and species names are proper names. Thus, like the names Fiordiligi and Dorabella, their primary function is to distinguish numerical individuals. Of course the two sisters from Ferrara (presumably) share a family name, indicative of the fact that they are related in terms of a common ancestor. However, individuals so related need not share intrinsic properties. Consequently, knowing what Fiordiligi looks like is of no help in the practical business of picking out Dorabella from all the other pretty women in Naples. As with the sisters, so with the species included in a genus.

Thus, in Hull's scheme, although the binomial has a theoretical significance or rationale, this significance does not carry over into the context of material practice.

Secondly, I wish to consider Hull's distinction between descriptions and definitions. From a realist perspective in the philosophy of science (see Bhaskar, 1978; Harré, 1986), lists of traits or properties produced in a purely empirical classificatory practice are, in effect, merely descriptions, since, prior to the elaboration of a theory which accounts for the possession of these properties by a particular or sample, we have no means of deciding which, if any, of these properties are necessary and which are contingent. But this state of affairs does not require us to conclude that the particulars with which we are dealing are parts of wholes rather than members of kinds. From a realist perspective, real definitions of kinds are not achieved at the initial taxonomic-descriptive stage of an investigation and they are not framed in empirical terms. The achievement of a real definition marks what is presumed, for the time being, to be the end of a phase of investigation, and a real definition, which is an identity statement, takes the form of a theoretical description of the 'microstructure' of the kind of entity concerned; that is, its nature or real essence. As regards the chemical elements, Hull (1975) seems to subscribe to a broadly similar view. Thus, in terms of current theory, gold is defined (theoretically) as an element with a particular atomic structure, more exactly a particular proton number, hence an atomic number of 79. But this theoretical description is, in principle, revisable or defeasible. Moreover, as I argued above, the term 'gold' must function in a practical as well as in a theoretical context. Now, statements can be produced in the practical context which refer to gold as a heavy, yellow and lustrous metal, but in current science, possession of these empirical properties would come fairly low down in the hierarchy of rules or criteria employed in the practical context for recognising a sample of material as a sample of gold (Harré, 1986). Indeed, in the case of one of them, colour, it would not be a reliable criterion in all circumstances for colour is a contingent property of gold; gold in mass is yellow but finely divided gold is purple or black. Statements about the colour of gold are, perhaps, more like "stereotypical" descriptive statements; gold is normally yellow. Somewhat higher in the current hierarchy of 'rules for recognition' of gold would be the possession by samples of dispositional properties such as specific melting point, boiling point and relative density, and rules employing these properties are higher because it is supposed that the possession of such properties is explicable, at least in principle or to some degree, in terms of the current theory of atomic structure. Would statements including such properties be analytic? In the strict sense arguably not, since the view that these properties are necessary properties is relative to the current theory of microstructure in terms of which they are (partially) grounded. Insofar as the theory of nature or real essence is revisable or defeasible, then so is the necessity of properties. Thus, unlike true definitions, statements in the practical context are only quasi-analytic (see Harré, 1986). It is evident from this account that statements which at one point in time have to be regarded as merely descriptive can become quasi-analytic and, conversely, at least in principle, quasi-analytic statements can become, or revert to, descriptive statements, these changes being consequent upon the current state of theory.

If this line of argument is accepted, Hull's observation concerning the descriptive nature of taxonomic characterisations must be regarded as inconclusive with regard to his overall argument. The rigid distinction between descriptions and definitions in the practical context is an aspect of the idealised, philosophical "high redefinition" of science (see Harré, 1986) which is inappropriate to the real material and cognitive practices of science. More significantly, the employment of empirical properties in the formulation of quasi-definitions – 'rules for recognition' in the practical context – is dependent upon the development of an explanatory theory which grounds these properties (or some of them) as, in principle, the necessary properties of a kind. In the absence of such a theory, the practical identification of individuals as members of kinds is indeterminate.

4

The Ontological Status of Taxa:
Theoretical Practice

"Under-labouring" for Natural Kinds

I argued in Chapter 3 that Hull's thesis does not provide a satisfactory basis for the practical business of assigning organisms to taxa. In practice, they are assigned to a particular taxon because of their intrinsic empirical properties, the morphological 'traits' or 'characteristics' which they actually possess and which can be distinguished in material practice. Further, I argued that if experimental practice is to be conceived as a rational practice, this presupposes that taxa are conceived as kinds. In sum, taxa are practically significant kinds. To complement their status in material practice we need a conception of taxa as theoretically significant kinds – natural kinds – because identification and classification can only become determinate if such morphological properties, or some of them, are conceived as necessary or essential to organisms of that kind. This entails the possibility of referring to common natures and requires a kind of explanatory theory in which properties are grounded in natures.

Now it must be conceded straightaway that, given the current state of biological theory, it is not possible to construct a compelling, positive argument for a theoretically grounded view of taxa as natural kinds. My role in this section, therefore, is to act as an "under-labourer" – though I appropriate Locke's expression with some trepidation since it suggests a possible comparison, in terms of quality of thought and argument, from which the present work is bound to emerge badly. Be this as it may, as "under-labourer," my task is that of clearing the ground so that the seeds of a revised natural kind conception may have some chance of germinating and developing. My strategy is, firstly, to attempt to undermine some of the arguments that have been advanced against the traditional view and, secondly, to outline the basis for a research programme in terms of which a new view akin to the traditional view might be constructed. This will involve developing a concept of the organism as a real structured thing with powers.

I begin, though, with a general consideration of the Theory of Descent and its explanatory power.

Narrative Explanations

It would be misleading to suggest that Darwinists in general and Hull in particular are unaware of many or most or even all of the problems I have discussed. Though not all of them have been directly addressed by Hull, it is clear from his published remarks that he does not regard either his general position or his demand for a "univocal species concept" as entirely devoid of problems; he is far from being dogmatic in this respect. Nevertheless, as I read him, I do not think it would be misleading to suppose that in the last analysis, Hull believes that such problems, or at any rate most of them, either can be resolved, or that strenuous attempts should be made to resolve them, in terms of the conceptual structure of evolutionary theory. And this belief could be regarded as justified to the extent that the view he is attacking, the idea that taxa are traditional classes, is itself fraught with problems and, more significantly, is not associated with any explanatory theory. Thus, insofar as it is possible, the relevant parts of biology should be rethought in terms of evolution theory; this is the thrust of his demand for a "univocal species concept." Hull (1976, p. 189) justifies this demand in terms of a belief in "the unity of science" and the corollary that "all scientific theories must be compatible."

Although I share Hull's view of the interdependence of classifications and theories, I have my doubts concerning the justification of research strategies or programmes in terms of a monistic view of science. The unification of science on anything other than a local scale remains a pious hope rather than an achievement. As regards the compatibility of theories, if this is taken as some kind of principle which is constitutive of knowledge and one branch of science or one scientific theory is treated as "well established," then we know from history (indeed the history of Darwinism) that the demand for immediate compatibility can be counterproductive. Compatibility is perhaps best regarded as a regulative principle; apparently incompatible theories should be treated critically and scrutinised carefully.

Now, if evolution theory is to be treated as foundational or unifying as Hull seems to suggest it should be treated, it is not unreasonable to demand

that it should be a theory which possesses considerable explanatory power
and scope in relation to central biological problems and in particular to the
problem of form, which is our central concern.

According to Hull (1978), "Evolutionary theory refers explicitly to organisms and species, not to Hitler and *Homo sapiens*" (p. 358). Thus in Hull's
evolutionary universe there are only two theoretically significant kinds, that
is, natural kinds: "species" (the species category) and organisms. There are
no natural kinds between these two 'levels of organisation' because particular
species taxa are conceptualised in the theory of descent as spatio-temporally
localised individuals, historical entities, and therefore have the same ontological status as individual organisms. Thus the explanation of any particular
instance of "descent with modification" must take a form analogous to the
biography of an individual, that is, the form of an historical narrative in which
particular species-individuals serve as "central subjects" (see Hull, 1975,
1984). Now, it is true that ideographic explanation is widely employed in
biology, ranging from Darwin's (1859, pp. 415–419) exemplary sketches of
"descent with modification" via the 'Just So' stories of the adaptationists
("So *that's* all right. . . . Do you see?") to the epic narratives of the palaeontologists. In this respect, therefore, the theoretical practice of evolutionary
biologists and their close kin is at least broadly consistent with Hull's claims
concerning the way in which taxa are conceptualised in the theory of descent.

For the purposes of this argument, I do not wish to dispute the general
claim that narrative accounts have explanatory power, though it is debatable
to what extent this power derives from overt or covert reference to the *kind*
of individual whose biography or history is being narrated. Rather, and in
the context of Hull's arguments, I wish to reemphasise the problem which,
as regards the explanation of morphology, these evolutionary narratives ignore or evade. The "central subjects" of historical narratives are individuals,
that is spatio-temporally continuous and cohesive entities. Napoleon, the
Kooh-i-Noor diamond and my Ford Fiesta are relatively unproblematic instances of such historical entities or continuants and in principle can serve
as the "central subjects" of narratives. Now we can certainly regard a species
taxon under the description, 'a set of organisms', as a continuant; offspring
are materially continuous with their parents. We can also regard a species
taxon under the description, 'a set of genes' (a gene pool) as a continuant;
to employ the current (if somewhat misleading) terminology, genes replicate.
However, if we regard a species taxon under the description actually employed by taxonomists and morphologists, 'a set of forms', a species taxon
is *not* a continuant, for there is a hiatus between parent and offspring with
respect to those empirical properties which constitute particular morpholo-

gies. There is no continuity between structures which are homologous in parent and offspring, for these structures are created anew in each individual. It is this hiatus which historical narratives evade or ignore (for a comparable argument, see Lévi-Strauss, 1968). It is therefore somewhat misleading of Hull (1981, p. 144) to refer to homologous "traits" along with species taxa (conceived as populations), organisms and genes as though they were all, ontologically, the same kind of thing and had the same criteria of identity. They are not and do not.

What is, in effect, a critique of the explanatory power of narratives of descent is given by Bateson (1886, 1894; see Webster, 1992). Firstly, Bateson (1886) argues that the search for the "common ancestor" with which to commence such narratives of descent is simply an evasion of the problem of form or specific organisation, for this ancestor is simply one more particular empirical form added to the world of particular forms and, as such, it has no special explanatory significance. Consequently, even if a "common ancestor" is found, its own empirical organisation stands in need of explanation and, therefore, this particular cannot serve as a means of explanation of the other particulars which are its descendants and share this organisation. In the first edition of the *Origin,* Darwin explicitly resorts to the Creator as a means of dealing with the problem of specific form or organisation, as Harriet Martineau noted with dismay (see Desmond and Moore, 1991), but in subsequent editions he is vague on the question. Since the phenomenon of specific form or empirical organisation is a *general* phenomenon, the explanation of organisation cannot be accomplished in purely historical terms by reference to particulars. Rather, as Bateson points out, explanation necessarily involves *scientific* knowledge which does not make reference to particular times and particular places, that is, knowledge of the "laws of growth and variation" (1886, p. 548). Bateson's point is identical to that made by Driesch (Chapter 1); the problem of understanding the link between the general and the particular, the 'type' and the individual is a nomothetic problem.

Bateson (1894) develops this argument in terms of the point I made earlier, namely, that the "central subjects" of those evolutionary histories which narrate the 'lives of forms' are imaginary not real individuals. The claim that the hypothesis of common descent accounts for the *distribution* of properties amongst individual organisms evidently rests on the assumption that there is a constant tendency for offspring to resemble their parents. However, no satisfactory theoretical account is provided in the *Origin* concerning the grounds for this assumption and the old theory of pangenesis which Darwin subsequently resurrected to provide this account, although broadly consistent with his position, happens to be false. Bateson's point regarding narrative explanations is made in a critique of the Darwinian concept of "inheritance," a concept which is based on the 'Ideal' interpretation of morphological correspondence discussed in Chapter 2. From this perspective, supposedly homologous morphological elements are effectively treated as though they are individuals, analogous to the items of private property which are transmitted from original proprietor to heir in human society and each of which has its individual and proper history. Because such narratives simply ignore the real hiatus between particular empirical forms, they tell us nothing about the means whereby such forms or their "modifications" are produced and consequently do not account for the possibility of their *existence.* As Bateson makes clear, such an account can only be given by means of a theory of a completely different type which addresses itself directly to the repetitive production of empirical forms, that is, a theory of morphogenesis. Since empirical forms are, in effect, created anew in each generation, Bateson argues that the notion of common origin must be replaced by the notion of common nature, which is to be understood as the productive or generative powers possessed by individual organisms by virtue of the kind to which they belong. It is in these terms that the "laws of growth and variation" are to be formulated.

Natures

With these points in mind, I turn to a consideration of Hull's claim that the Darwinist conception of species taxa as historical entities or individuals necessitates a particular view of their theoretical status, viz, that they are not "theoretically significant *kinds*" (Hull, 1976, p. 189), hence not natural kinds. Hull (1978) argues that current theoretical practice in biology is consistent with this view.

To talk of natural kinds is to talk of natures which are common to members of these kinds. However, it appears that Hull will brook no talk of common natures in relation to taxa. In the context of a discussion of the possibility of a science of particular species, for example, the human species, he argues that:

If species are interpreted as historical entities, then particular organisms belong in a particular species because they are part of that genealogical nexus, not because they possess any essential traits. No species has an essence in this sense. Hence there is no such thing as human nature. There may be characteristics which all and only extant human beings possess, but this state of affairs is contingent, depending on the current evolutionary state of *Homo sapiens.* Just as not all crows are black (even potentially), it may well be the case that not all people are rational (even potentially). (Hull, 1978, p. 358)

As I read it, two distinct claims seem to be involved here. The first is a reiteration of the Kripkean thesis of the necessity of origin but the contingency of development. That is the conceptual claim that, in the light of the ontological status of taxa within Darwinian theory, the properties of a species do not provide criteria for identity; they are contingent rather than necessary or essential with respect to identity so that a species taxon qua individual can, in principle, change its properties indefinitely and remain one and the same taxon. The second is an empirical claim that, in the light of the same theory, the properties of a species taxon are as a matter of fact historically contingent. Hull seems to suggest that the conceptual claim and the empirical claim are congruent and, on this basis, he rejects the idea of a common nature or essence which is necessarily possessed by all the individuals which comprise a particular species.

Now, it is not at issue that the empirical properties of the individual organisms which are supposed to be members (or 'parts') of a particular taxon actually vary as a matter of fact in the present. And from this fact it is not unreasonable to infer that they have varied in the past and will vary in the future; in this sense, they are indisputably contingent, "depending on the current evolutionary state." But little can be concluded from this fact as regards the question of the existence or nonexistence of common natures. The answer to this question hinges upon how this variation is to be understood, that is, explained, and, consequently, what more precise sense we are to give to the notion of contingency as regards matters of fact. Are we to understand the variable forms in terms of the 'typical' properties or characteristics of a kind, which implies the notion of *some* constraint on what is possible consequent upon the nature of the kind in question? In this case, properties will be conceived as relatively contingent; as will be explained in detail subsequently, they are contingent insofar as the circumstantial factors involved in their manifestation are contingent, but are necessary given these factors and the specific nature of the kind in question. Alternatively, are we to suppose that properties are absolutely contingent, that there are no common natures and therefore anything is possible by way of properties and their combinations? The latter view seems to be the one which has been espoused by the Darwinist tradition and is apparently espoused by Hull since he rejects talk about common natures.

From the perspective of "Common Sense," *Homo sapiens* and *Cygnus olor* are the names of kinds of things (continuants) and properties cluster because they are the 'typical' or necessary properties of things of that kind.

From a Darwinist perspective, such names are merely the names of absolutely contingent aggregates of independent properties; "plastic conglomerates of miscellaneous attributes" (Bateson, 1894, p. 80)) or "frozen accidents" (Resnik, 1994). If properties are independent then so are predicates and, in principle, the only constraint on possible predicates is the formal constraint embodied in the principle of noncontradiction (Harré and Madden, 1975). Hence, whatever statement can be made about, say, swans, never amounts to more than an accidental generalisation. At present, not all swans are white though all have beaks. In the past perhaps all swans were white with beaks. In the future it may be the case once more than all swans will be white though without webbed feet. Or it may be the case that not all future swans will have feathers or beaks or webbed feet, or that some or all future swans will look like creatures from the nightmares of Hieronymous Bosch, or like what we now call snails or rabbits or tulips. And why not? As Nietzsche might have said, if essential natures are eliminated, everything is permitted.

It is doubtful whether any biologists have, in practice, actually subscribed to views of this fantastical kind, nor, I suggest, are there any good reasons why they should do so. From a Darwinist perspective, taxonomy loses its fundamental scientific status and becomes a relatively trivial (and, I have suggested, indeterminate) pursuit of the actual. In the course of a discussion of the view proposed by the "Darwinian Phylogenists" that forms are 'accidental', Driesch (1914) observes that

the totality of living forms . . . appeared to them as meaningless as, say, the forms of clouds in their accidental peculiarity. But this at once did away with any deeper meaning for zoological classification. It was settled once and for all; the question had no sense. (p. 141)

Yet many taxonomists continue to regard their disagreements as significant, which only makes sense if the enterprise is regarded as an enquiry (albeit preliminary) into the real, determinate and systematic structure of nature. The existence of empirical variation within a supposed species taxon does not appear to be conceived by all taxonomists as indicating the absolute contingency of properties, for it often results in debates as to whether, in the context of material practice, we are dealing with more than one taxon. Should a set of individual plants comprising the section Bulbocodium of the genus *Narcissus* be regarded as members of one very variable species – *Narcissus bulbocodium* – or are we dealing with five (or even more) distinct taxa? (Blanchard, 1990). Insofar as such debates hinge upon the intrinsic properties of individuals, they presuppose the reality of kinds and the 'typical' nature of such properties. Moreover, as the preceding example suggests, many or most accounts of the classificatory problems arising from experienced specific variation presuppose experienced relative invariance at a generic level. In this sense, many organisms apparently reproduce 'after their kind' in a relatively unproblematic fashion. Thus Darwin (1859) may have trouble distinguishing the various species of bramble in Britain, but at least he can recognise that they are all brambles. Baker recognises 16 species of *Narcissus* and therefore disagrees with Haworth who recognises 150 species (Blanchard, 1990). Nevertheless, they agree about the object of their disagreement, the specification of the (relatively) unproblematic genus *Narcissus*. Hull (1978) draws attention to our experience of the great dissimilarity in properties that can exist between members of a taxon, for example, the males and females of some species or the different castes of social insects. But these are not "one-off," unique variations; they occur regularly and repeatedly within a taxon, as do the markedly dissimilar forms that comprise all life cycles. Many of the more dramatic empirical irregularities that exist within a species taxon are, if an oxymoron be permitted, regular irregularities, as Bateson (1894) argues. Experienced variation of this sort may support an argument that taxa are not empirical classes in the traditional sense, but it does not necessarily support an argument that they are not kinds in some other sense. And it gives little support to the view that forms are absolutely contingent and that there are no common natures. Empirical regularity (even of 'irregularities') is, after all, at least prima facie evidence for the existence of law. Bateson summarises the conclusions from his empirical study of variation thus:

The crude belief that living beings are plastic conglomerates of miscellaneous attributes, and that order of form or Symmetry have been impressed upon this medley by Selection alone; and that by Variation any of these attributes may be subtracted or any other attribute added in indefinite proportion, is a fancy which the Study of Variation does not support. (Bateson, 1894, p. 80)

However, as I previously emphasised, purely empirical considerations drawn from material practice can be no more than suggestive as regards the status of properties and the existence of natural kinds. The issue can only be decided by means of explanatory theory, as indeed Bateson, for one, recognises. My claim is that there is no compelling, that is, theoretical, reason for asserting the absolute contingency of properties for the simple reason that, currently, there is no adequate explanatory theory.

The theory which Hull invokes, the Theory of Descent, at least in its "classical" form, has nothing relevant to say, in the context of explanation, regarding the status of empirical properties. The theory claims that individual organisms possess most of the properties they do because their ancestors possessed these properties; as Darwin (1859) puts it "the chief part of the organisation of every being is simply due to inheritance" from a "common progenitor" (p. 228). As Bateson (1886) points out, and as I noted earlier, the theory does not *explain* the existence of empirical morphologies but simply takes them as given in the common ancestor. Furthermore, Bateson (1894, 1913) notes that empirical variation between individuals is also taken as given by Darwinian theory. Thus variant forms are the given "modifications" of the given; they are brute facts, unanalysed and unexplained. In terms of the theory, there is no necessity for either stability or variation in the empirical properties of the individuals that comprise a species taxon, that is, a temporally extended population in Darwinian terms. All that is claimed is that *if* variant forms occur and *if* they they are adaptively advantageous then they will be accumulated within the population as a consequence of natural selection. There is no *explanatory* basis here for making any claims whatsoever regarding either the necessity or the contingency of properties considered as matters of fact.

Bateson's own view is that at least some variant forms are "typical" in the sense in which that term is used by Driesch and other students of *Entwicklungsmechanik;* that is, they are characteristic manifestations of a kind. Thus, *Crocus* flowers with threefold and with fourfold rotational symmetry would both be "typical" manifestations of a generative mechanism characteristic of a kind – they are alternative "Positions of Organic Stability" – though whether this kind corresponds to the empirical taxon Iridaceae is an open question. For Bateson, the explanation of variation is simply one aspect of the causal explanation of form in general; the task is to understand the "essential" aspects of diversity in terms of an explanatory theory (Bateson, 1913). I will examine below the modifications of "classical" Darwinism consequent upon the assimilation of some aspects of genetics. Before attempting this, however, it is necessary to clarify the argument via an examination of a realist view of the nature of scientific laws and explanatory theory.

Natural Kinds as Theoretically Significant Kinds

The concept of a natural kind cannot be equated with the concept of a mere class for any collection of entities having at least one characteristic in common can constitute a class; the "proverbial rocks in a box" (Hull, 1981, p. 150) are members of a class but they are not members of a natural kind. A preliminary definition of a natural kind term is that it is a term which plays a certain kind of role in a scientific theory (Putnam, 1975); hence natural kinds are theoretically significant kinds. As part of his argument against species taxa being natural kinds, Hull (1978) argues that species taxa are not conceived by biologists as theoretically significant kinds. I quote him in full.

If species are actually spatio-temporally unrestricted classes, then they are the sorts of things which can function in laws. "All swans are white," if true, might be a law of nature, and generations of philosophers have treated it as such. If statements of the form "species *X* has the property *Y*" were actually laws of nature, one might rightly expect biologists to be disturbed when they are proven false. To the contrary, biologists expect exceptions to exist. At any one time, a particular percentage of a species of crows will be non-black. No one expects this percentage to be universal or to remain fixed. Species may be classes, but they are not very important classes because their names function in no scientific laws. Given the traditional analysis of scientific laws, statements which refer to particular species do not count as scientific laws, as they should not if species are spatio-temporally localised individuals. (p. 353)

The thrust of this argument seems to be that biologists do not regard any empirical generalisations concerning particular species as scientific laws because they are unperturbed by their falsification. The existence of black swans, for example, causes no sleepless nights. Since the names of species taxa do not actually function in scientific laws within biological discourse, species taxa are not, and are not regarded as, theoretically significant or natural kinds. Rather, this state of affairs is consistent with species taxa actually being individuals and being regarded as such by biologists.

Hull's argument here seems to be conducted in terms of a positivist theory of science. From a positivist perspective, the theoretical part of science consists of a formal system of statements organised according to the canons of deductive logic. Among such statements are scientific laws which, in the "traditional analysis" referred to by Hull, are conceived of as statements concerning empirical regularities of either coexistence or succession – that is, 'universal' or 'covering' laws. From this perspective, "All swans are white" might be such a law and, as such, its falsification would be significant. On this view, explanation is regarded as the deduction of a description of events or states of affairs from a statement of the appropriate covering law together with a statement of the initial conditions and, indeed, elsewhere, Hull (1976, p. 189) defines explanation as "subsuming a particular instance under a law of nature." Thus, from a positivist perspective, a deductive system of statements is at the heart of a theory and any associated picture of mechanisms is regarded as of merely heuristic value. However, it has been argued that this positivist notion of science is inadequate on several counts (see Harré, 1970; Harré and Madden, 1975) and should be replaced by a realist view of science which inverts the traditional order of things. From a realist perspective, models of mechanisms occupy the central position in theory and the achievement of a deductively organised system of statements is a desirable heuristic device. Here I consider only some aspects of this argument.

Within the positivist tradition a genuine law of nature is conceived as being more than a mere generalisation, that is, more than a mere summary of instance statistics, for a genuine law is accorded a special kind of generality which endows it with superior logical power such that it can be used to infer unexamined cases. Consequently, a central problem within this tradition is that of distinguishing genuine laws – "nomic universal statements" – from merely accidental generalisations, since the latter are also supported by instances. Much effort has been devoted, and many gallons of ink expended, on the task of attempting to make this distinction in the only terms this tradition recognizes, that is, purely formal terms (see Harré and Madden, 1975). One formal, intrinsic characteristic which has been proposed to distinguish lawlike statements from merely accidental universal statements is that the former should be "unrestricted universals." From this perspective, the statement, "All the panels on Webster's car are rusty," while true, would be merely an accidental universal since the particulars of which the predicate "rusty" is affirmed (the panels) are restricted in that they are spatio-temporally localised by virtue of the fact that they are parts of a spatio-temporally localised particular (my car). By contrast, genuinely lawlike statements are supposed to be spatio-temporally unrestricted generalisations (see Hull, 1978, p. 337); that is, the scope of predication is not restricted to objects which fall into a particular spatial region or a particular period of time. Hull's claim that, from the "traditional" perspective, statements of the form "species X has property Y" are not laws of nature (and are not regarded as such by biologists) is based on the view that a genuine law of nature must be an "unrestricted universal" combined with the fact that, in the Theory of Descent, species taxa are conceptualised as individuals, that is, spatio-temporally located particulars. From this it follows that, even if, say, the statement "All swans are white" happened to be true, it would not be a law of nature but merely an accidental universal statement. Hence the imperturbability of biologists in the face of counterinstances is both intelligible and justified.

Two points can be made in this connection; one concerns the relevance of evolution theory, the second concerns the positivist conception of theory employed in the argument. If accepted, these points render Hull's conclusions problematic.

I argued earlier in this chapter that the Theory of Descent does not provide an adequate *explanation* of the morphological properties of individuals. If this argument is correct, then the fact that species taxa are conceptualised in this theory as spatio-temporally restricted entities is not relevant to the business of deciding the status – lawful or accidental – of any general statements which might be formulated concerning the morphological properties of individuals which are members of such taxa.

The second point, concerning the positivist conception of theory, is more fundamental in the context of the present discussion. Harré and Madden (1975) argue persuasively, and in detail, that the possession of formal characteristics like "unrestricted universality" is, at best, only a necessary condition for a statement to be lawlike; it is not sufficient. They further claim that, in fact, "there is no set of necessary and sufficient formal conditions which a universal statement must meet before it can be called a 'nomic universal' " (p. 32). They go on to argue that the only sure way of distinguishing lawful and accidental universal statements is by seeing why, in the first case, the observed regularity must hold; that is, giving reasons for its necessity. Thus we regard the regularity in the behaviour of gases expressed in, say, Boyle's Law as a genuine law (though "Ideal") because, given the kinetic theory of gases, we see why the relations expressed in this law must obtain; the 'must' here is an expression of explanatory necessity. If the nature of gases, that is, their constitution, is in reality as it is modelled in the theory, then gases necessarily behave in the manner described by the gas laws. Hence, according to Harré and Madden, the criterion for accepting a statement as lawlike is that we can attach the statement to a theory which provides a picture of the hypothetical generative mechanism responsible for producing whatever patterns and regularities are observed in nature and expressed in the instance statistics. Thus the construction of hypothetical models of generative mechanisms is central to the theoretical practice of science, and this is one reason, though by no means the only one, why Harré (1970) argues that the traditional positivist philosophy of science should be replaced by a realist philosophy which proposes a different theory of theory. From this perspective a natural kind is conceived as a set of entities having the same constitution or nature. However, natural kinds do not necessarily manifest themselves as classes in the sense that all members possess a determinate set of manifest or actual characteristic in common; not all samples of water are liquid, not all carbon is black and not all alligators are male. Nevertheless, the *total set* of manifest and latent characteristics of a thing, insofar as the thing is conceived as a member of a natural kind, is not arbitrary or accidental but is causally connected with the inner constitution of the thing in question, with its real essence. It is this connection which justifies classification in terms of manifested characteristics (Harré, 1970, p. 197).

I believe that, just as M Jourdain had been speaking prose all his life without realising it, so biologists, or at least experimental biologists, have a realist rather than a positivist view of science whether they are aware of it or not for, on the whole, they see their central theoretical task as the elaboration of models of mechanisms rather than the formulation of universal laws which state simple empirical invariances such as "All swans are white." With this task in mind, although they look initially for *some* kind of pattern in the natural phenomena, this does not have to be a strict empirical invariance, though it may be. Thus, for example, if it turned out, on initial examination, that all crows in a natural population were either black, white or grey, then the situation is such that some attempt at explanation might be thought worthwhile. On the other hand, if it turned out that crows could be any colour, or any combination of colours, I doubt that biologists would take any further interest.

Thus I believe that Hull's claim that biologists "expect exceptions to exist," or at least are often unperturbed by their existence, is correct, though not for the reasons he suggests and only in a qualified sense. A qualification is necessary since the claim is only correct if we restrict ourselves to a consideration of naturally occurring phenomena. Experimental biologists, at least in principle, expect to discover some exceptionless regularities in that they attempt to *produce* them by manipulating situational conditions. In this they are at least sometimes successful, though the practical isolation of a single causal mechanism and the closure of the system to unwanted extraneous influences can be extremely difficult to achieve, as anyone who has worked with living organisms will know.

Biologists also indicate their tacit adherence to realism by their preference for explanatory theories which refer to the nature of the entities concerned. They do not, on the whole, conceive explanation as subsuming an instance under a law of nature. And, given the poverty-stricken nature of the deductivist conception of explanation (see Harré, 1970; also von Wright, 1971), they are justified in this view. Question: Why is this bird black? Answer: Because this bird is a crow. Question: But why are crows black? Answer: All crows are black. Any further questioning can only result in reiteration of this sequence and our demand for an explanation remains unsatisfied. If we are to be given an explanation of *why* all crows are black (or, indeed, why only some crows are black) we need to be told something about the nature of crows. What it is about crows that "gives rise to" the colour; what characteristic other than colour do birds of this kind possess which causes the colour of their plumage; how is it possible for their feathers to be coloured? In other words we want an explanatory theory of feather pigmentation in crows. But, in terms of a deductivist conception of explanation taken strictly, it is gratuitous to seek further than the covering law since the deduction of the particular case from the law has been achieved (Harré, 1970, p. 21).

The general position of biologists is well captured by Woodger (1948) when he notes that they do not accord the same status to all empirical generalisations. Thus he observes that the generalisation ''All swans are white'' is merely empirical – a summary of instance statistics – and involves no explanatory theory from which it follows that all swans are necessarily white. Consequently, its falsification by the discovery of a black swan leaves biologists unperturbed. However, the generalisation ''All adult mammals have a heart'' is not simply empirical but is related to a physiological (functional) theory which claims that organisms of the size and complexity of mammals *must* have a heart. Hence, Woodger suggests, the falsification of this generalisation would have profound consequences for biology for it would be the falsification of a (supposed) genuine law pertaining to large, complex animals, of which mammals are an instance. More significantly, it would imply that an important explanatory theory was false.

In sum, the question as to whether species and other taxa can be conceived as theoretically significant kinds, hence natural kinds, can be answered only by appealing to a kind of explanatory theory in which the members of such taxa are conceived as possessing a common constitution or nature which is responsible for their morphological characteristics. No such theory exists at present though it is the aim of this book to outline a possible theory.

The Causal Explanation of Manifest Properties

With these points in mind, let us consider the properties of an undisputed natural kind such as water. I want to argue, from a realist perspective, that there is no contradiction between regarding ''states of affairs'' – the possession of particular, manifest properties – as relatively contingent and referring to laws and to natures or real essences.

The empirical properties of samples of water are variable. Not all naturally occurring samples of water are liquid, nor for that matter do all samples of water boil at the same temperature. It is contingent whether a particular sample of water is a solid, a liquid or a gas since the possession of any one of these properties is dependent upon the contingent circumstances of temperature and pressure in which the sample finds itself. Or better, the system in which it finds itself, since a kettle on a gas ring in a particular geographical situation constitutes a system.

Now, it cannot be part of the nature of water to be solid, liquid and gaseous for these states are mutually exclusive. Nevertheless, the possession, in relation to circumstances or systemic conditions, of any one of these properties by a sample is explicable in terms of the nature (the real essence) of water, that is, in terms of current scientific theory, its molecular constitution. It is the aim of explanatory theory to explain how and why, when the temperature rises to 100°C as a result of the application of heat, and under closed conditions as regards pressure, a sample of water exhibits specific or 'typical' properties; why it boils rather than turning solid. Particular 'states of affairs' are explained in terms of intrinsic dispositions, which are dependent upon natures, and situational conditions, which are contingent. Natures have to be understood as determining a repertoire of natural possibilities, what samples of a kind can do, that is have the power to do. To say that water has the power to boil is to say that a sample of water will boil in the appropriate conditions *in virtue of its intrinsic nature.* (see Harré and Madden, 1975, p. 86). Insofar as it is dependent upon circumstances, what the sample actually does, is a contingent matter; that is, properties or behaviour can be regarded as relatively contingent. However, in a particular, determinate circumstance, that is, under closed conditions (Bhaskar, 1978), only one possibility, can be actualised, and sole possibility is equivalent to necessity. In terms of current scientific theory, the nature of water is such that a sample of it cannot turn solid or red or give a spirited rendering of the Hallelujah chorus when the temperature reaches 100°C at normal atmospheric pressure; rather, it must boil. Thus explanatory theory, by grounding certain dispositional properties or modes of behaviour in natures, permits such properties or behaviours to be regarded as 'typical' or 'essential' – part of the nominal essence of the kind – such that they can be used as criteria of identity. A sample of a liquid that does not boil under the specified situational conditions is not a sample of the kind 'water' but of some other kind.

If the actualisation of possibilities is dependent upon contingent circumstances, it follows that the 'typical' properties of a kind such as water need not be universal properties in the sense that they are *actually* manifested by every sample of the kind since it need not be the case that every sample of the kind actually finds itself in the appropriate circumstances; not all samples of water are liquid. The 'typical', empirical properties of water which are of interest to science are dispositional properties. Consequently, if there are to be universal, lawlike statements which feature the names of kinds such as water, these cannot take the form of descriptions of simple sensory qualities manifested here and now in all samples. Rather, they must take the form of conditional ("If . . . then . . .") statements referring to what a sample of the kind will do in the appropriate circumstances. Such statements, taken together with a categorical statement referring to the nature (the real essence) of the kind by virtue of which it possesses these dispositional properties, amount to a statement of the *power* of the kind in question (see Harré, 1970; Harré and Madden, 1975). To ascribe a nature to a thing is to say what that thing *is*. In terms of current scientific theory, the nature or real essence of water is its molecular constitution, hence water is necessarily H_2O and anything which does not have this molecular constitution, or (possibly) one which can be regarded, on theoretical grounds, as a variant of it, is not water. The explication of the concept 'water' as H_2O provides the 'real definition' of 'water'.

Understanding the nature of a (kind of) thing as an enduring generative mechanism places a natural constraint on, or reduces the field of the possible with regard to properties. Properties (or some of them) do not just happen to cluster accidentally so that they can be gained or lost independently; rather, they belong together precisely because they are the properties of a kind of thing in the sense that they are produced by the same causal mechanism. Consequently, from a realist perspective, predicates are not independent. For example, if hydrogen is as a matter of fact composed of atoms of identical structure, and if this atomic structure accounts both for the valency and for the spectrum of the element, then it is not possible that the element has a certain spectrum but does not have certain chemical properties and vice versa; the two coexist necessarily (Harré and Madden, 1975).

In the light of these considerations, a central question, to put the matter as simply as possible, is whether, and if so, to what extent, we can talk about organisms and their forms in a broadly analogous fashion to the way we talk about kettles of water on gas rings. That is, can we talk realistically about the variable manifest properties of members of a taxon in the way we talk realistically about the variable empirical properties of water in terms of intrinsic dispositions, powers and natures, on the one hand, and situational conditions on the other? On the face of it, it would appear that we can talk in this way in some cases and up to a point.

Taxa are not classes in the sense that all members possess a determinate set of manifest characteristics; not all members of species *X* actually manifest property *Y* and the same seems to be true of higher taxa. However, not only is it the case that not all swans are white, it is also the case that, at an early stage of development, *no* individual swans are white. Along with all the other morphological properties of adult swans, the colour of their plumage does not becomes manifest until a particular stage of development. Hence, the manifest properties of adult organisms must be regarded as kinds of dispositional properties. Consequently, if there are to be any universal, lawlike statements relating to the morphological properties of a taxon, conceived as a natural kind, they must take a conditional form something like: "If members of taxon X are subject to stimulus G then they will manifest property Y."

The determination of sex provides an example of this kind since, in a number of species, contingent environmental factors determine whether an individual develops male or female morphological characteristics (see Goodwin, 1991 for an introductory review). Thus, for example, in the marine worm *Bonellia* the male lives as a parasite inside the female. If a fertilised egg falls onto the proboscis of a female, it develops into a male; otherwise it develops into a female. Some fish are like Tiresias and during the course of their lives change sex according to environmental circumstances. The best analogue of the kettle on the gas ring is provided by species of alligators where it can be shown experimentally that, if the temperature is below 30°C almost all embryos develop into females, if above 34°C almost all develop into males; thus, we can *experimentally* produce a close approximation to an exceptionless regularity as regards sexual morphology. Evidently, any individual alligator embryo has the potential to develop into either male or female; hence it is plausible to suppose that all crocodiles have a common, bipotential constitution with respect to sexual morphology. What actually happens in a given individual is dependent upon situational conditions. Since a difference in temperature causes a difference in morphology, the morphological sex of alligators is dependent upon circumstances and in this sense is contingent. However, the mode of response of alligators to these circumstances seems to be *specific* to them since, as far as we know, no other kinds of organisms respond to environmental stimuli of this type in precisely this way, that is, produce the set of sexual morphologies characteristic of alligators. Hence a complete explanation must also take account of the nature of alligators and the dispositions, that is the powers, which they possess consequent upon this nature; powers which, apparently, are not possessed by daffodils, hummingbirds or human beings.

From a realist perspective alligator embryos appear to be 'powerful particulars' (Harré and Madden, 1975). What an individual alligator actually does is to be explained in terms of what alligators can do – the actual is explained as an instance of the possible – and what alligators can do is a consequence of the kinds of things they are and the powers they possess as a consequence of being a thing of that kind.

Must we conceive of a distinct power for each kind of manifest morphology? Harré argues that science uses the notion of power in such a way that powers are related to determinables but not to determinates;

A material thing has the power to be coloured, not *the* power to be red. Changing the circumstances, say by varying the colour of the incident light, changes the determinate colour a thing looks, but it is the very same power which in some circumstances is manifested as red and in another as black. (1970, p. 312)

If this is the case with respect to morphology, we could say that each individual alligator embryo, by virtue of its nature, has a power which can be characterised, in an abstract way, as the power to produce a 'alligator sexual morphology'. This power is manifested as male or female morphologies depending on the situational conditions in which the embryo finds itself. Such a view is consistent with the field theory of morphogenesis developed by Goodwin in this book. It is also consistent with the common sense view that the males and females of a species are variants of a single natural kind. Moreover, it is the case that, in many organisms, male and female sexual morphologies can be related in terms of morphological correspondence (homology), hence conceived as relatively simple empirical transformations of each other.

It would seem then, that, up to a point, we have here some kind of a parallel with the case of water and its properties and behaviour. This parallel provides some grounds for regarding the sexual morphogenesis of alligators as a paradigm for morphogenesis in general and, as I will argue, there are some grounds for claiming some kind of causal equivalence between external circumstances or situational conditions and internal conditions or stimuli such as genetic constitution. However, it is evident that the parallel with water is incomplete. On the basis of the observations, we can formulate conditional statements (laws) relating the sexual characteristics of alligators to environmental circumstances but these are merely empirical generalisations, formally comparable to the (pretheoretical) generalisations we can make about water. Consequently, they could be regarded as merely accidental universal statements. Although we can outline the *kind* of explanatory scheme which is required, as I have done above, there is, at present, no *actual* explanatory theory of morphogenesis which would enable us to claim that the behavior of alligators is *necessary* in virtue of the intrinsic nature which all individuals of the kind possess in the way that we can claim that the behavior of samples of water is necessary. Thus we have no grounds for claiming that these modes of behavior are part of the nominal essence of 'alligator', hence that they constitute criteria of identity for members of that kind. To be able to make this claim we need a theory of the structural basis of the powers we have invoked; we need to cash out the phrase "in virtue of their intrinsic nature" in terms of a causal or generative mechanism. Here new problems arise.

Individual Essences

The notion of explaining the manifest properties of individuals in terms of intrinsic natures and situational conditions is certainly not foreign to the Darwinist tradition. Indeed, Hull himself expounds a view of this kind in the course of a discussion of Freud and the notion of "normal psychological development" (Hull, 1984). He explains,

Species commonly possess the ability to proceed down alternative developmental pathways given the appropriate sequence of conditions. These pathways taken collectively are termed the 'reaction norm' for the species. Certain pathways may be quite common because the requisite environmental conditions are quite common, but all developmental pathways are equally 'normal' in the sense of lying within the reaction norm for the species. (p. 38)

In this context he cites one of the examples I have used above, *Bonellia,* whose larval form is described as possessing the 'potential' of being either male or female; for 'potential', I would wish to substitute 'power'.

Now, it would appear that Hull, in common with most scholars in the Darwinist tradition, equates situational conditions with external, environmental conditions. And, from earlier remarks in the same paper, it would appear that he conceives 'potential' (or powers) in terms of "genetic information" and this is presumably also the case with respect to the colour of crows mentioned previously, "Not all crows are black (even potentially)." In a vigorous polemic against the idea of species-specific essences he contends that:

The amount of intellectual wriggling that goes on to make *Homo sapiens* monothetic is dismaying. 'Human beings are essentially *X*. Any human being who ostensibly lacks *X* really possesses it at least potentially. So a particular person lacks the genetic information for the formation of thumbs. That person still has a thumb at least potentially. Conversely, the exhibition of *X* by members of any other species is only apparent. . . . I find such special pleading intellectually disgusting. (Hull, 1984, p. 35).

This conception of 'potentials' or powers in terms of "genetic information" is, of course, the conventional wisdom of the neo-Darwinist tradition, a synthesis of Darwin's Theory of Descent and Weismann's conception of the organism together with the assimilation of the discoveries of classical, Mendelian genetics to the Weismannist paradigm. It has virtually become the folk wisdom of the western world. The received view is spelled out by Hull (1976) in the course of a discussion of "individual essences":

Every organism . . . possesses a genetic makeup which not only remains largely unchanged during the course of its ontogenetic development but also *directs* this development. Thus the genetic constitution of an organism might be viewed as its "individual essence." In this sense, having largely the same genetic makeup is a necessary condition for an organism remaining the same individual; it is not sufficient. Spatiotemporal unity and continuity are also necessary. (p. 177, emphasis added).

Given our current knowledge of genetics (see subsequent discussion), I think a plausible case could be made for regarding the genetic constitution of an individual as that individual's essence and a necessary condition of its numerical identity. In other words, the possession of a particular, complex, genetic constitution gives that individual the particular, idiosyncratic individuality that it has and accounts for its difference from other individuals with which it might be confused, apart, that is, from its monozygotic twin, which is why additional properties are necessary. Now, from the perspective of Darwinian theory a species taxon is a temporally extended population of individuals. In the light of current knowledge, we have no reason to suppose that the individual organisms which comprise a species taxon necessarily have an identical genetic makeup; indeed, we sometimes have positive knowledge that they do not. Thus, if "genetic makeup" or "genetic information" is the major component of individual essence or nature, we have no reason to suppose that any one individual essence is necessarily identical to any other individual essence, hence no reason to suppose the existence within a species of a *common* essence or nature; the most we can expect is some kind of family resemblance between the accidental clusters of essences. The 'gene pool' of neo-Darwinian theory is effectively the sum of the genetic constitutions of all the individuals which comprise the population; it is an historical entity and, in principle, any changes are possible. Consequently, to reapply Hull's own words, while there may be a genetic constitution which is possessed by all and only the extant individuals which constitute a taxon, this state of affairs is contingent and depends on the current evolutionary state of the taxon. Hence, insofar as we can talk about species-specific 'potentials', powers or 'reaction norms' these are nothing more than accidental generalisations about a particular taxon. As Hull (1984) puts it, such talk is "a description of how a particular system functions for a certain period under particular circumstances" (p. 39). The view that the 'essence' of a species taxon (insofar as it makes sense to employ such a term) is effectively the 'sum' of individual essences is a position which is consistent with Hull's overall thesis that species taxa are not natural kinds. The same argument applies to 'higher' taxa with, if anything, even more force. I note, in passing, that, with 10^4 to 10^5 genes per organism, sometimes more, it does not seem particularly plausible to suppose that natural kind distinctions can be made on the basis of relative degrees of similarity and difference in genetic makeup (cf. Kitts and Kitts, 1979).

To speak of natural kinds is to speak of qualitative identity and, consequently, to treat individuals as atoms, all exactly alike. Hence, any attempt to save the notion of taxa as natural kinds necessitates the development of a concept of the real essence or nature of the kind which is distinct from the essence of the individual, conceived as accidental genetic constitution. I will attempt to develop such a concept via an analysis and critique of the claim that the "genetic makeup" of the organism "directs" its development. Here, I will argue that, at least in some circumstances, "genetic makeup" must be conceived as a situational condition or, more exactly, an efficient cause of form and, further, that the nature or real essence of a kind should be equated with the material cause of form which is identified with a morphogenetic field.

The Weismannist Paradigm

In the course of a discussion of the "concept of organism," Woodger (1930) points out the persistent tendency within biology to accept Descartes' view of the organism as some kind of machine, that is, an entity in which the relations between the parts are external or nonconstitutive. As Woodger notes, the idea of a machine without a transcendent mechanic or 'organising principle' is absurd and, for Descartes, the mechanic is God. The claim that the essence of an organism is to be found in the nuclear genome can be seen as a legacy of this idea in its late-nineteenth-century formulation.

Horses tend to beget horses. It was Weismann (1883) who correctly argued that the phenomenon of "heredity" must be understood as the repetitive *production* of 'similar' forms. Hence, the 'similarity' of forms must be understood in terms of morphogenesis. In discussions of morphogenesis current at the time, it was assumed that the process must be analysed atomistically in terms of the intrinsic properties of cells and Weismann (1885) supposes that the properties of a cell depend only on its nucleus. Hence the "chromatin" which determines the properties of cells must be different in each kind of cell and he proposes that this comes about by the qualitative division of the structurally complex nucleus of the egg into simpler nuclei, each set of which stands in causal relation to a part of the adult organism. Consequently, as he puts it, "the essence of heredity is the transmission of a nuclear substance of specific structure" (p. 65). Thus, in the conceptual scheme of Weismann, "Chromatin takes the place of Descartes' God as the 'controlling' mechanic" (Woodger, 1930, p. 18) and this view lives on in those interpretations of the theory of the gene in which the nuclear genome or "genetic information" or a "genetic programme" is conceived as "directing," "controlling" or "instructing" the process of individual development. A recent expression of this view can be found in an outline of the rationale of the Human Genome Project:

'[The] collection of chromosomes in the fertilized egg constitutes the complete set of instructions for development, determining the timing and details of the formation of the heart, the central nervous system, the immune system, and every other organ and tissue required for life.' (Delisi, 1988)

The Weismannist position attains its *reductio ad absurdam* in the fantasies of sociobiology in which the organism is regarded as merely a vehicle for genes.

Now, it is clear that vague expressions such as 'directs', 'controls' or 'instructs' are attempts to fill a theoretical vacuum by means of metaphor (see Woodger, 1945). Moreover, in the context of discussions of organismic morphology, the use of the concept "genetic information" is equally metaphorical. Whereas in the context of molecular biology the concept of "information" has a precise significance consequent upon its role in explanatory theories concerned with the production of proteins of specific structure, it has no such significance in relation to the production of, say, thumbs, to use Hull's example, because there is no comparable theory. In this context, to speak of "genetic information for thumbs" is merely to employ a kind of metaphorical shorthand to summarise the fact that, as regards organisms which can interbreed, there is a genetic difference between organisms which have a particular property and those which do not, and the genetic difference can be causally correlated with the morphological difference. The conceptual and theoretical gap between explaining the structure of a protein and explaining the structure of the hand is comparable to the gap which, according to Lévi-Strauss (1969), some people remarked in Grétry's harmony: "Between his high notes and his low you could drive a carriage."

However, the problem does not lie solely in the metaphorical nature of concepts such as 'directs' and 'controls' but more crucially in the misleadingly one-sided fashion in which the metaphors are employed. Weismann's concept of "nuclear preformation" with its associated hypothesis of the qualitative division of nuclei was disproved by Driesch's famous experiments involving the disruption of cleavage patterns (see Driesch, 1908). Consequently, the idea that the "essence of heredity" lies in the nucleus has no foundation. As Hull himself observes, current knowledge suggests that the "genetic makeup" of all cells (apart from the gametes) in a multicellular organism is qualitatively identical and remains identical throughout the life of the organism. Therefore, if we wish to continue to speak metaphorically, we are obliged to invert the original claim and speak of the structure of the organism and the course of development as "controlling" or "directing" the "expression of genetic information." We might be better advised to abandon vague metaphors and attempt to develop an explanatory theory.

As Woodger (1945; see also Bateson, 1913, 1928) points out, the classical "Mendelian" Theory of the Gene was designed to explain certain kinds of differences between individual organisms and, by extension, the distribution of properties among the individual members of interbreeding populations; that is, differences between organisms which still resemble each other sufficiently closely to be regarded "kinds of fowls" or "kinds of sweet peas" (Bateson, 1928, p. 279). It was not designed to explain the course of development – morphogenesis – which involves temporally and spatially organised changes in the properties of individual organisms. If "genetic information" is qualitatively the same at all times in all cells of a multicellular organism, then the differences which make the temporal and spatial differences cannot be genetic differences. Thus, explanation of morphogenesis requires a theory of a different, and a more comprehensive, kind than the Theory of the Gene. Such a theory must be designed to account not only for the findings of genetics but also for those properties of the organism revealed by experimental embryology and experimental morphology. Let us leave this problem on one side for the moment and pursue the Theory of the Gene.

The theory, in the form in which it is usually taught at an elementary level and the form in which it has entered Western folk consciousness, claims that in certain cases where two organisms differ, there is a genetic, that is, a chromosomal, difference that makes the differences. The theory, therefore, necessarily presupposes that the organism can be regarded as a "closed system" (Bhaskar, 1978), for to claim that it is a *genetic* difference that makes the difference is precisely to claim that the system is closed to all other causal stimuli which might, in principle, be involved. The organism is closed either in the sense that it is actually isolated from such stimuli or that their action can be regarded as constant in the situation which is of interest. Woodger (1930) makes essentially the same point in a slightly different way. A closed system is thus one in which a constant conjunction of events obtains. Now, it seems but a short step from the presupposition of closure in relation to the explanation of certain kinds of morphological differences to the assumption of the truth of the metaphysical thesis of "regularity determinism" (Bhaskar, 1978). This thesis, which can be summarised in terms of the canonical formula, "whenever X, then Y," asserts that the cause is both necessary and sufficient for the effect, hence that the same cause has the same effect and the same effect is always due to the same cause. In biological terms, this amounts to the claim that the presence of a specific gene or set of genes is a necessary and sufficient condition of the possibility of the production of a specific morphological character; hence, in the absence of the gene[s], the organism does not have the power to produce the character in question, for example, the thumbs cited by Hull. This position is consistent with an 'atomistic' conception of the organism of the kind advocated by one of the early Mendelians, de Vries (1900) who argues that the traditional concept of the species should be replaced by one which regarded the species as "a composite of independent factors . . . the units of species specific traits are to be seen in this connection as sharply separate entities" and further, "the total character of a plant is built up of distinct units. These . . . elements of the species . . . its elementary characters, are conceived of as tied to bearers of matter, a special form of material bearer corresponding to each individual character." Although de Vries is hostile to Darwinism, his views are echoed by one of the architects of the "Synthetic Theory," Fisher (1936), who claims that "the living organism is an aggregation of characters in the form of units of some description." This view tends to be implicitly conveyed in elementary courses in genetics and is captured in folk expressions like "the gene for eyes." Such 'atomism' is consistent with the Darwinist notion that properties and, consequently, also parts, are independent and just happen to coexist in a particular organism as a consequence of the current evolutionary state of the gene pool. Thus, the palaeontologist Osborn (1915) asserts: "A principle of *hereditary separability* whereby the body is a colony, a mosaic, of single, individual and separable characters . . . each character has its independent and separate history" (p. 198, 194).

Genes and Organisms

On the basis of the findings of classical "Mendelian" genetics we can claim with confidence that, in some cases at least, there is a genetic – chromosomal – difference that makes – causes – a difference in morphological properties between two individuals. There is a constant conjunction of events. But this is, at best, only a partial explanation of the manifestation of properties. For we do not, except in the case of relatively simple properties like colour, know *how* the difference is made, nor why it is *this* difference rather than some other difference. The point is brought home when we recognise that, for the most part, we do not understand how it comes about that, sometimes, a genetic difference makes *no* morphological difference, nor why the difference that is made depends upon the '*material*' upon which a gene acts, nor how it happens that an environmental or experimentally imposed difference can sometimes *mimic* the morphological effect of a genetic difference. In other words, while we do have real knowledge of *some* efficient causes or causal stimuli involved in the production of forms, we have, at present, little in the way of detailed knowledge of the *causal mechanisms* involved and, therefore, scant knowledge of how any particular form is *possible* nor, consequently, of whether or not there are inherent constraints on the possible. From a realist perspective, the real essences of members of a kind are to be identified with common causal mechanisms.

As Waddington (1956, p. 5) observes, the problem of gene activity "is essentially an embryological problem" and it is in relation to the causal mechanisms involved in the production of forms that geneticists primarily concerned with development, such as Waddington (1940, 1956, 1957) and Goldschmidt (1958), have attempted to formulate an 'organismic' conception of gene action. In so doing, they have given some preliminary indications as to how we might begin to move beyond the Weismannist idea that the specific essence of the organism is to be identified with the genome.

Goldschmidt (1958) emphasises that it is a central tenet of genetics that a mutant locus or the normal allele does not "control a character" but is only a differential; the production of a 'character' involves many genes, if not all of them. As Goldschmidt puts it: "If we wish to express this factual situation by saying that a phenotypic trait is the product of action of many or all genes, we must realise that this *façon à parler* is nothing but a circumscription, in terms of the atomistic theory of the gene, of the fact of the unity and integrity of the organism'' (p. 250). Waddington (1957) argues in a similar fashion. Now, if the production of a 'character' involves many genes it follows that a single gene may be involved in the production of more than one 'character' and, indeed, such pleiotropic effects are well known. To give one example, in *Drosophila* the gene *morula* causes abnormalities in the number of cell divisions in the facet-forming cells of the eye and also causes a reduction in the size of the bristles in bristle-forming cells (Waddington, 1956). Thus, the morphological effects of a particular gene depend upon the 'material' upon which the gene acts. Consequently, at the level of morphology, while the formula 'whenever X, then Y' is sometimes true, it is not generally true.

This conclusion is supported more directly by examples of mutations such as *eyeless* which, initially, results in a failure of eye development in *Drosophila* (see Harland, 1936). Here, the causal factor is a recessive gene and if a stock of individuals homozygous for *eyeless* is inbred for several generations then flies eventually appear which possess normal eyes. Such flies are still genetically *eyeless,* as can be shown by back-crossing procedures. The conventional, if obscure, explanation of this phenomenon is to suppose that there has been a "reorganisation" of the gene complex so as to prevent the morphological (phenotypic) expression of the *eyeless* allele. Irrespective of the correct explanation, the example shows that a specific genetic makeup is neither necessary nor sufficient for the production of a specific morphology. Whatever may be the case with regard to presence or absence of thumbs cited by Hull, we can say that, in *Drosophila,* the power (Hull's 'potential') to produce eyes is not dependent upon the presence of specific "genetic information" in the sense of a chromosomal locus of a specific nature. The simple-minded might be inclined to wonder whether eyes should not be regarded as 'typical' forms that a particular kind of *thing* – a particular imaginal disc of *Drosophila* – has the power to produce by virtue of its intrinsic nature, hence will produce in the appropriate circumstances, or even produce spontaneously.

The example just given shows that there can be invariance of morphology despite variation in genetic makeup; there are no constant conjunctions here. There is also evidence which supports this conclusion by means of the converse argument, namely that invariant genetic makeup does not necessarily result in morphological invariance. As Goldschmidt (1958) observes, there are variable, 'quantitative' aspects to mutation in addition to qualitative aspects. Thus, there are variations in the extent to which any particular gene is 'expressed' in terms of morphology within a population and within any given individual. There are also internal (genetic) and external environmental effects on 'expression' of a 'quantitative' nature. From such observations Goldschmidt concludes that

if the quantitative element of mutation is so sensitive to external and internal environments, it might follow that all or most mutational effects take place within a spectrum of possible developmental changes the range of which alone determines the possibility of appearance of *both* mutational or environmental effects. (1958, p. 257, emphasis added)

In other words, to gloss Goldschmidt's conclusion in terms of the concepts employed in this discussion, the specificity of the effect of a causal stimulus – the change it brings about in an individual – whether this stimulus be internal or external, is to be understood in terms of what the *organism* – a continuant – has the specific power to do; that is, what it can do by virtue of its specific nature – the kind of organism it is – and not solely in terms of the specific nature of the stimulus itself. We should, perhaps, think of internal or external stimuli as providing the occasion, the required situational conditions, for the exercise of a power. Similar, 'organismic' conclusions are drawn by Waddington (1940, 1956, 1957).

It is in this context, as Goldschmidt (1958) emphasises, that the phenomenon of "phenocopying" acquires significance. Needham (1942) notes that in many organisms, changes in morphology – phenotype – brought about by genetic changes (known or supposed) may be imitated by external factors or may occur apparently spontaneously, as in certain kinds of terata. Such effects can be distinguished from genetically caused effects only by the fact that they are not reproduced in any offspring. Some of the earliest systematic observations of this kind were made at the end of the last century when it was found that the wing pattern of butterflies can be changed by temperature shocks so that the experimentally induced pattern is indistinguishable from the pattern characteristic of other races of the same species. However, the real importance of such phenomena was only appreciated in 1917 by Goldschmidt.

For Goldschmidt (1958), phenocopying is "the basic phenomenon for any study of genic action" (p. 262). He claims that the morphological characteristics – the phenotypes – of *Drosophila* mutants can be exactly copied by relatively nonspecific perturbations, such as temperature shocks, applied at a critical period in development. This is also true of some other organisms. Goldschmidt summarises the facts thus:

In *Drosophila* . . . every known type of mutant can . . . be copied as a phenocopy . . . practically any kind of shock treatment produces phenocopies . . . many unspecific shocks at the proper time can produce the same phenocopy, but also . . . one and the same shock may produce different phenocopies. (pp. 258–259)

We might note the apparent analogy between phenocopying and the fact (discussed earlier) that in some species sexual differences are environmentally caused while in others the difference is genetically caused. We might further note the apparent similarity between these phenomena and what Bateson (1894) calls "homoeotic variation," which is particularly common in arthropods and flowering plants, where, within a single individual, and consequent upon genetic perturbation, one member of a meristic series assumes the properties which are normally characteristic of another member; for example, legs are produced instead of antennae (see Ouweneel, 1976). Moreover, homoeosis seems to be but a developmental form of heteromorphosis (see Needham,

1952) which is relatively common when organisms regenerate, say, an appendage, subsequent to experimental or accidental loss and where there are no grounds for claiming any change in genetic composition; for example, the regeneration of a chela instead of a maxillipede by a crab. It is in this context that we might wish to situate observations of the kind that, whereas both the cockroach and the *Drosophila* limb normally has a five-jointed tarsus, the regenerate in the cockroach is four-jointed, which, significantly, is typical for the taxon Locustodea, as Bateson (1894) emphasises, and a four-jointed form can be produced genetically in *Drosophila* (Waddington, 1940).

Thus, apparently equivalent variations of form can be produced within a single individual or in different individuals either spontaneously (as it seems) or as a response to *either* environmental *or* genetic perturbations. And we might wish to follow Waddington (1940) and draw parallels between these sorts of phenomena and those revealed directly by experimental embryology; for example, the fact that in *Rana fusca* the development of a lens from the ectoderm requires the presence of an extrinsic causal stimulus from the underlying eye cup, which acts as an "organiser," whereas in *Rana esculenta* the lens can develop spontaneously in the absence of an eye cup. As Waddington explains:

> The similarity between the theoretical schemes we have arrived at on embryological and genetic grounds is ... apparent. In embryonic development we are confronted with alternative modes of development, the choice between which is taken in reference to an external stimulus, in inductive development, or to an internal one, in the case of mosaic development. In considering the effects of genes, we find alternatives the choice between which may be taken in response to diffusible substances, as in *Drosophila* eye colours, or apparently in response to internal factors as in aristopedia. It is clear that we have merely followed two different methods of approach to the same phenomena, and that the two schemes are in fact identical. (p. 91)

Waddington attempts to give a pictorial analogy for this state of affairs in the "Epigenetic Landscape" which is, in effect, a temporalised version of Bateson's (1894) concept of "Positions of Organic Stability," itself derived from Galton (1889). Indeed, Waddington's argument is, effectively, a development of Bateson's (1894) claim that "the system of an organised being is such that the result of its disturbance may be *specific*" (p. 74, emphasis added). That is to say, the result of disturbance is 'typical' or characteristic of a kind of being. A similar claim was made by E. Geoffroy St Hilaire at the beginning of the nineteenth century (see Russell, 1916). The claim suggests that the response, in terms of the form produced, of a biological system to perturbation – the action of a particular causal stimulus – is determined by its intrinsic powers, that is, the inherent nature of the particular causal mechanism implicated. Waddington's (1940, 1957) concept of "competence" is effectively a concept of power and his choice of the term "evocator" for the causal stimulus is apt, given its connotation of 'calling forth'. These powers, while dependent upon the presence of a nuclear genome of some kind (as demonstrated by enucleation experiments), seem, at least in some cases where the matter can be directly investigated, not to be dependent upon the presence of specific "genetic information," if by this is meant the presence of a particular chromosomal locus or set of loci of a specific nature; the latter is neither necessary nor sufficient for the production of a specific form. Organisms are not closed systems and the canonical formula 'whenever X, then Y' – where X is a particular gene or set of genes – does not always obtain. Moreover, the observations on phenocopies and related phenomena suggests an equivalence between particular genetic and environmental factors as efficient causes of difference in form; both kinds of contingent factors seem to provide the situational conditions or occasions for the exercise of powers inherent in the organism and its parts. In sum, the evidence does not provide support for the Weismannist view, espoused by Hull and the Darwinist tradition, that the nuclear genome is the real essence of the organism considered as the member of a kind. Rather, it suggests that the organism and its parts must themselves be conceived as 'powerful particulars'. Consequently, if the real essence is equated with the causal mechanism responsible for the production of manifest morphological characteristics, this essence must be sought in the morphogenetically relevant structure of the organism and its parts.

Paradigms of Action

In the preceding discussions I have, from time to time, invoked or employed the concept of 'power'. It is, perhaps, useful to explicate the concept further in relation to the present concerns.

It is a self-evident truth that the 'production' of eyes, thumbs and other morphological features is a consequence and expression of changes in the properties of organisms and their parts, that is, changes in complex material beings. This truth is so self-evident that it is apt to be taken as a truism that is hardly worthy of discussion. Consequently, in the rush to explain, the significance of this fact is often ignored. The most striking aspect of the Weismannist, mechanistic tradition, in the context of the problem of morphogenesis, is the 'disappearance of the organism' since the theory proposes, on the one hand, an organising 'centre' consisting of a set of independent, contingently variable causal 'factors'; on the other, a set of independent morphological 'characters' which are the effects of these 'causes'. All causes are regarded as efficient. Evidently, Weismannism has a certain resemblance to classical behaviourism. Just as radical behaviourism eliminates 'mind' or any notion of 'cognitive structure' and attempts to explain everything in terms of extrinsic causal stimuli and behavioral responses related to each other in temporal succession (here too 'everything is possible'), so Weismannism eliminates the organism, conceived as a real, complex and specific kind of material structure. Weismann's organisms are structureless and in this sense have disappeared.

The legacy of Weismann's distinction between an 'active' germ plasm and a 'passive' somatoplasm is a continuing tendency in some quarters to view, if only implicitly, the nuclear genome as a causally 'active' centre and the developmental process as merely a sequence of events which happen to 'passive' materials. This tendency is reflected in the figurative language of much current biological discourse: the 'direction' or 'control' of development by the genome; in the context of morphogenesis, "genetic information" on the one hand, mere matter or material waiting to be 'informed' on the other; genes are frequently 'activated', organisms rarely. As I noted earlier, the views of both Waddington and Goldschmidt can be interpreted as implying that the effects of genes (and environmental factors) have to be understood in terms of the powers of continuants, the organism and its parts. But, further to these specifically biological arguments, there are more general reasons for calling into question atomistic and mechanistic conceptions of explanation.

As Harré and Madden (1975, p. 82; see also Bhaskar, 1978) note, there is, within the mechanical 'world view' a general tendency to think of events as happenings to passive things and to conceive all causes as efficient and extrinsic to the thing which changes. The central paradigm of action here is the colliding billiard balls of mechanics. Here it is supposed that the balls have no intrinsic, active powers of motion though they can be moved. The final state of the system is the product of an extrinsic factor, the (variable) states of motion of the balls before collision, and an intrinsic factor, the (constant) inertia or mass of a ball. On this model, all action is to be understood as the production of an effect by an impressed or stimulating cause, the intrinsic factor being wholly passive.

However, as Harré and Madden point out, there are many other paradigms of action to choose from. At the opposite pole to that of mechanics is the paradigm of spontaneous, voluntary human action in which the source of the action is intrinsic to the actor so that explanation is couched in terms of the actor's powers. In between the two poles of things which are wholly patients and things which are wholly agents can be found a spectrum of paradigms which differ in the proportion of explanatory weight which is assigned, on the one hand, to intrinsic factors such as constitution or nature, and on the other, to extrinsic factors such as the stimuli to which things respond. A stick of dynamite, perhaps, lies somewhere in the middle of the spectrum since its power to explode, although dependent upon intrinsic states, is exercised only in the appropriate circumstances, whilst an extremely unstable explosive, such as ammonium triiodide, lies towards the agent end since it is liable to go off spontaneously. The fact that no naturally occurring things can be regarded as pure patients (Harré, 1970) and that it is at least debatable whether anything can be regarded as a pure agent does not undermine the significance of the basic point.

Harré and Madden note: "There has seemed to be something fishy and soft, occult and mysterious about the second [agent] paradigm, and something tough, scientific and empirical about the first [mechanical]" (p. 83). Consequently, it is not surprising that some philosophers and natural scientists have preferred the first paradigm and have regarded the second, with its concept of power, as a residue of magic and appealing only to the tender-minded, to those unable to "face the stern truths of empiricism" with its requirement that the "concept of essential nature . . . be replaced by that of the accidental collocation of qualities." From this empiricist perspective, a claim such as that made by the doctors in Molière's *Le Malade imaginaire* that opium possesses a "dormitive virtue" is regarded as the epitome of a vacuous pseudoexplanation. However, as Harré and Madden argue, this is simply "mistaken metaphysics" (p. 84), for the concept of power is neither occult nor magical. Hence, contrary to the empiricist position, to invoke the "dormitive virtue" of opium is to make the first move in developing a scientific explanation of how the substance has the effects that it does.

The result of this mistaken metaphysics has been the dominance of the mechanical paradigm until relatively recently, together with the associated belief that explanations should be couched primarily in terms of extrinsic, causal stimuli or efficient causes. As Bhaskar (1978) observes, this position is exemplified in the assertion by Hobbes that, as regards change, "nothing taketh a beginning from itself." But, as Bhaskar points out, if this is interpreted as meaning that it takes *no* part of its beginning from itself, if none of the total set of necessary and sufficient conditions for an event (effect) is intrinsic to the thing or material, then this is tantamount to assuming that the event will occur whenever the antecedent cause is present whatever the thing or material. In other words, we assume 'Whenever X, then Y.' Consequently, we should expect to hear the sound of breaking glass whatever object the stone strikes, to observe a red colouration whatever kind of paper is dipped into an acid and to observe the growth of a leg on any part of any organism which possesses the gene *aristopaedia*. If, however, there are substance or material conditions for the event which are such that, if they were not satisfied, the event would not occur, then these conditions must be included as part of the 'total cause' (in J. S. Mill's sense). In order to hear a characteristic 'crash and tinkle' rather than a 'dull thud' when a stone hits an object, certain intrinsic, material conditions are necessary, viz, that the object be made of glass; that is made of a material which has a certain nature (structure) and, consequently, certain passive powers (liabilities) which are described by means of the predicate 'brittle'.

For Harré and Madden, an important aspect of the concept of power is that it can incorporate a strong sense of potentiality – what would happen as a matter of course if interfering conditions were absent. Consequently, efficient causes are to be understood not only as the presence of stimuli which activate a quiescent individual but also as the absence or removal of constraints upon an already active individual. Moreover, as Bhaskar points out, just as not all causes are efficient causes, so not all efficient causes are extrinsic; it is pure prejudice to suppose that they must be such. Thus, the structure of a magnetic field may be the efficient cause of what happens in it, though a material cause is required as well – materials must be magnetisable. In general, the properties of a whole may explain the properties of its component parts, as the organisation of society – the structure of social relations – explains (in part) the properties of persons.

Harré and Madden and Bhaskar argue that to deny that certain intrinsic, material conditions are necessary for the occurrence of a particular event is, in effect, to deny the principle of material continuity. And this is to deny that change can be rationally explained, for the principle maintains that events are always *changes* in things, not the *replacement* of one thing by a completely new thing. We think of the world as consisting of things – continuants – which endure through some, but not all, changes. And we think of events as changes in things. The thing which is changed is already given as structured and complex. If it is partially changed, material continuity is preserved through change: the window is broken but the glass preserves continuity; the litmus paper turns red but remains intact (for a time); the chemical atom enters into a new combination but preserves continuity through the chemical reaction. Thus, in all explanation of change a material cause, a continuant, is required in addition to an efficient cause. The mechanistic paradigm directs attention almost exclusively to the latter. However, as Bhaskar observes, in many sciences the most important developments have resulted from the search for 'hidden' things or continuants resulting in the identification of new natural kinds, and this search has often been provoked by the prima facie appearance of creation ex nihilo.

Morphogenetic Fields

Individual organisms are materially continuous entities, and morphogenesis, manifestly, involves changes in the qualitative characteristics or modes of behavior of *given* material. However, the morphogenetic process presents the prima facie appearance of ex nihilo creation. For, while the given material is certainly structured in the way that all living material is structured, it is nevertheless true that, with rare exceptions, no visible spatial organisation can be found in this material which can be related to the complex spatial forms and patterns which are produced; as we say, development is epigenesis. The task, therefore, is to discover the relevant 'hidden' structures which must exist if morphogenesis is to be rationally explained. It was the students of experimental morphology and embryology in the early decades of this century who set themselves this task and what they discovered was a new kind of entity, the morphogenetic field.

It is not my intention in this section to attempt a complete account of the morphogenetic field, a topic of considerable complexity. Rather, I will restrict my discussion to a brief exposition of the concept of the field and attempt to outline some aspects of its possible significance in relation to the present concerns. The classical work on fields is discussed by Gurwitsch (1921), Woodger (1930), Bertalanffy and Woodger (1933), Huxley and De Beer (1934), Spemann (1938), Weiss (1939), Waddington (1940) and Needham (1942); some of the most significant work subsequent to the second world war is reviewed by Waddington (1956, 1962). More recent views are fully discussed by Goodwin in the second part of this book.

As with so much else that is theoretically significant in relation to a putative science of form, the field concept has its origin in the work of Driesch, though he did not elaborate the concept. Its source lies in his experimental demonstration that, contrary to the Roux–Weismann hypothesis, each cell of a sea urchin embryo, when isolated at the two-cell stage, does not produce a half-embryo but a complete, miniature pluteus larva of normal form. Subsequently, he demonstrated that whole normal larvae could be produced from single cells isolated at the four-cell stage, from eggs whose cleavage patterns had been altered (as remarked earlier), from two eggs fused together and from a single egg divided into two along the meridional plane (see Driesch, 1908). The sea-urchin embryo, therefore, possesses powers of self-regulation such that a *normal* form will be produced following a variety of drastic experimental perturbations. Work subsequent to that of Driesch demonstrated that these powers, though considerable, are not, as he originally thought, unlimited. As Hörstadius demonstrated, if the the sea urchin egg or embryo is divided equatorially, only the vegetal pole produces a normal embryo whilst the animal pole is incapable of self-regulation. Other investigators subsequently demonstrated experimentally that the powers of self-regulation discovered by Driesch in the sea urchin embryo also existed in individuals of many other species at some stage of their development and, in the case of organisms which can regenerate lost parts, for example hydroids and some amphibians, throughout their lives. They are, perhaps, universal properties of organisms, including noncellular organisms.

The experiments referred to involve the isolation and transplantation of the material of developing organisms – that is, changing the spatial relations of the material. Under such conditions, and if the experiments are performed at the appropriate time, the normal developmental fate (the prospective significance) of the material is altered and it develops in a manner which is consistent with its new relations. The results of experiments of this type are often summarised in the aphorism: "Developmental fate is a function of position." Thus the material acquires or exercises (or loses or fails to exercise) the power (or liability) to change qualitatively, that is, to produce specific morphological structures, as a result of the spatial relations it enters into. Since entities related only spatially are not connected, the fact that the material undergoes changes in properties following an alteration of its spatial relations can only be explained if we assume the existence in the developing organism of a 'hidden', spatially organised system of constitutive relations. As Woodger (1930) points out, the fact that the 'parts' of organisms are internally or constitutively related distinguishes organisms from machines, whose 'parts' are only related externally; the latter remain unchanged when their relations are changed.

The denotation of these spatially organised systems of constitutive relations by the term 'Field' is due to Gurwitsch and Weiss who drew attention to some apparent formal resemblances between these systems and magnetic fields – a resemblance in part noted, but not developed, by Driesch. In particular, the power of the morphogenetic field to cause changes in the properties of undetermined material transplanted into it so that this material develops in a manner appropriate to the field it is in; the power of the field to maintain its properties when reduced in mass or divided into two; the power of the field to restore its unity when fused with another field of the same kind. Whether these resemblances between physical fields and morpho-genetic fields are merely superficial or indicate some real relationship is an open question.

Much of the classical experimental investigation of field phenomena was carried out in various species of amphibians, though broadly comparable results have been obtained in other species. Up to gastrulation the embryo behaves as a unitary field, a single system of constitutive relations, and the developmental fate of a part is a function of its position in the whole. In this respect, it is comparable to the early sea urchin embryo or to an adult organism such as *Hydra* (see Webster, 1971) which possesses considerable powers of self-regulation and behaves as a unitary whole. However, with the onset of neurulation, secondary fields appear, sometimes called individuation fields. At this point, the embryo has, to a greater or lesser extent, lost its unitary character and has become a spatially organised mosaic of relatively autonomous regions, each region corresponding to a future part or organ system; fore and hind limbs, heart, neural tube, and so forth. If these regions are removed, the embryo will not usually restore them by self-regulation. There are at least eleven distinguishable pairs of these regions in the amphibian embryo (Needham, 1942) and each of these regions possesses a more or less fixed and determinate nature of a 'generic' kind, for example, as presumptive fore limb or hind limb. Regions retain their natures when transplanted and will produce the morphological structures that they would have produced if left in situ. Thus, although it must be assumed that the nature of each region has been caused by its original position (relations) within the whole which is the primary field, this nature is now determined as an intrinsic property of the region. However, each of these regions is itself a field and possesses powers of self-regulation, so the 'specific' parts of the organ or organ system are undetermined. For example, the heart is normally formed by the fusion of a pair of fields, but if this fusion is prevented, each field will produce a miniature but normal heart. Likewise, a single limb bud can produce three complete limbs when divided and indifferent material transplanted into the early limb bud will be assimilated. However, since the nature of the material within a particular field is 'generically' determined, the result of transplantion experiments are a function of position and nature. Thus, for example, in the chick embryo, material transplanted from the proximal part of the hind limb to the distal part of the fore limb produces hind limb structures in accordance with its 'generic' nature but these structure are those appropriate or 'specific' to its new, distal, position in the fore limb (Saunders, Cairns and Gasseling, 1957).

Thus the process of development in the amphibian embryo, and in the embryos of species which are broadly similar, involves the production of parts within a whole – a field – and these parts are themselves wholes – fields – within which further parts are produced. The hypothesis proposed in this book is that morphogenetic fields are the material causes of morphogenesis; they are, in Harré and Madden's terminology, 'powerful particulars', and different kinds of morphological structures are the empirical manifestations of their different powers which in turn are dependent upon their natures – what they *are*. The hypothesis proposed here is that the nature or essence of a field is its structure, and this structure must be conceived as dynamic to account for the phenomenon of self-regulation. As 'powerful particulars', fields are at least candidates for the status of natural kinds of things.

Fields are "wholes actively organising themselves" (Needham, 1942, p. 129). Hence, with respect to at least this aspect of their behaviour, it appears that fields must be conceived as possessed of active powers; they are agents not patients, and if not pure agents, then certainly close to the agent end of the spectrum of paradigms of action. As I noted earlier, although the field concept can be said to have its source in the work of Driesch, he himself did not elaborate it and his views developed in a different direction. The full reasons for this are too various and complex to be discussed here, but a primary factor seems to have been his commitment to a mechanistic conception of the organism coupled with his awareness that a machine requires a mechanic, a requirement of which many of his contemporaries remained cheerfully oblivious; Bateson, to a limited degree, was one exception (see his various remarks concerning the atomism of genetics in Bateson, 1928). From a perspective of atomism and mechanism, the question of form presents itself as a *quite distinct problem of order* and the problem becomes acute in the light of the facts of self-regulation; hence Driesch (1908) emphasises "the importance of an arranging and ruling factor in spite of all units" (p. 292). Since Weismann's concept of 'God in the Genome' has been rejected on empirical grounds, Driesch turns elsewhere for the required "arranging factor" and revives the Aristotelian concept of "Entelechy" which is, in the first instance, "a mere word" used to "signify . . . all that happens in morphogenesis with respect to order" (p. 104). Entelechy denotes the "invariable factor of wholeness" which manifests itself in the phenomenon of self-regulation and which indicates that there is "at work a something which bears the end in itself" (p. 106). As is well known, this "something" acquires a central explanatory role in Driesch's 'philosophy of the organism' and apparently ends up as an agent which exists outside the space-time-causal system which is Nature, though it acts into space (p. 254). Driesch thus explicitly (and self-consciously) departs from the Kantianism which is his avowed start-

ing point. Whether his final position is coherent is debatable, but this notwithstanding, it is clear that while entelechy may (or may not) be a successful
metaphysical postulate, it cannot, by definition, be an object of scientific
knowledge since it is not an object in Nature.

Now, as Woodger (1930) emphasises, the problems to which Driesch
draws attention are very real problems, though, as he also notes, they are
problems which, at least to some extent, arise from mechanistic and atomistic
presuppositions. If morphogenetic fields exist and if they are conceived as
unitary systems of internal, that is, constitutive, relations and if these systems
possess intrinsic powers of self-organisation by virtue of their nature, then
we are no longer working within a mechanistic paradigm and it is arguable
that we no longer require a concept of a distinct "arranging factor". We do,
however, need a theory of the nature of fields which accounts for the intrinsic
powers of self-organisation. As regards a "something which bears the end
in itself"; following Harré and Madden, I have noted above that the non-
mechanistic concept of active power, or agency, incorporates a strong sense
of potentiality, of what would happen 'spontaneously' if interfering conditions were absent. From this perspective, if the end of ammonium triiodide
is to explode, then a sample of that substance bears this end in itself since it
has the *power* to explode by virtue of its nature (what it *is*), and it *will*
explode unless it is prevented from doing so. Remarks of this kind are merely
pointers to future theoretical research since, at present, there is no completely
adequate *theory* of fields in terms of which they could be justified (see Goodwin in Part II). However, as regards the *existence* of morphogenetic fields, I
think there is little room for debate. Hacking (1983, pp. 23, 36) has argued
that the term 'electron' denotes something real, quite irrespective of the truth
or falsity of theories about electrons, because we can "spray" electrons. In
a particular material practice, the term 'electron' is used to refer to whatever
it is that causes a change in the charge of a niobium droplet; hence we can
speak of the reality of electrons on the basis of their specific causal powers.
Likewise, the term 'morphogenetic field' denotes *something* real, irrespective
of the adequacy of our theories, because we can physically manipulate fields
and because they cause things to happen. Knowledge of the existence of fields
is, arguably, a permanent addition to our knowledge of what exists in the
world.

I think that the reality of a 'something' would be admitted by all competent
students of morphogenesis. Where they differ is in their theoretical accounts
of this 'something'. Wolpert (1969, 1991; see Goodwin in Part II), for example, tends to give a rather minimalist account of fields in the sense that,
in his explanatory scheme of the generation of forms, the greatest explanatory

weight is given to (contingent) "genetic information." There is little basis here for any conception of taxa as natural kinds. By contrast, from Goodwin's perspective, the greatest explanatory weight is given to the structure of the field which is conceived as a 'powerful particular'. Consequently, Goodwin's scheme provides a much more promising basis for any view of forms as members of natural kinds.

If, as Goodwin argues, the essence or nature of a field is its dynamic structure, then it is in terms of this structure that we must account both for its powers to produce particular morphological structures and for any constraints there may be on its possible behaviour in terms of the kinds of structures it can produce. Thus, it is in terms of a theory of fields as the causal mechanisms of forms that we must explain the empirical variations of forms and understand their necessity. This issue is discussed in detail by Goodwin, but a brief summary is in order here. In terms of the theoretical model which Goodwin develops, fields are conceived as dynamical systems and genetic or environmental factors are supposed to determine parametric values in the equations which describe the structure of the field. Such factors therefore act to 'select' and stabilise one empirical form from the set of forms which are possible for that type of field; in effect, these factors act to close the system. Goodwin further argues that morphogenetic fields exhibit what he calls "robust" properties. What this amounts to is that the same manifest form can be produced over a range of parameter variations. Thus there will be a tendency for some forms to be statistically frequent, therefore, 'normal' while others are infrequent, therefore 'abnormal'. However, all these forms are 'typical' of the kind in the sense that they are all produced by the same kind of field; that is, the set of fields can be described by the same set of equations.

It is impressive that the prescient Bateson (1913, 1928; see Webster, 1992) envisaged a theory of this general type when he speculated that the notion of "Positions of Organic Stability" might be explicated in terms of a dynamical system and drew an analogy between organic forms and Chladni figures – the patterns produced in suitable media when excited by sound. In such systems, variation in, for example, the frequency of excitation while other factors are held constant – shape and size of plate, amplitude of sound and so on, results in the production of a family of patterns which resemble each other while differing in the number of individual elements, that is, nodal lines (see Jenny, 1967, 1974).

If morphogenetic fields are the material causes of morphogenesis – the 'elementary' causal mechanisms responsible for particular morphological structures – then, from a realist perspective, it is in terms of fields that we have to think of real essences of kinds. Fields appear to be systems of relations and I have suggested that their nature, that is their real essence, lies in their dynamic structure. Thus in the case of the amphibian embryo, and others like it, where there is a single, primary field, a single 'powerful particular', which is responsible for the production of 'basic' parts, it would seem that we can speak of a single nature or real essence. However, insofar as these 'basic' parts are themselves relatively autonomous fields, apparently possessed of distinct and different natures, and which are responsible for the production of 'parts within parts', then we seem obliged to speak of multiple real essences. As I noted previously, there seem to be at least eleven different pairs of fields in the amphibian embryo, each with distinct powers. Thus, while the adult individual conceived at a rather general level – a whole with parts – might be regarded as an 'element', the same individual, conceived at a more specific level in which the parts consist of parts, should, perhaps, be regarded as a 'compound'. As I noted in Chapter 1, Bhaskar (1978) has suggested that, insofar as the 'ordinary things' of the world can be regarded as compounds, then, in principle, we can only expect them to resemble each other empirically. If this argument is correct, then, if the same basic field can produce different, though related, morphologies in different circumstances, as Goodwin argues, we should only expect empirical resemblances between adult individual organisms when these organisms are described in detail, that is, specifically.

To speak of the sameness of fields and the resemblance between individuals is to raise once more the question of classification. I take it as established that empirical organisms cannot be conceived as members of traditional classes in the strict sense. I also assume that, if taxa are to be reliable, they must distinguish natural kinds (Harré, 1986); that is, the classification must be of causal mechanisms or real essences. Now, if a logic of traditional classes cannot serve as a means for systematically representing the diversity of forms, we need to examine the possibility of using a different mode of representation, one based on a logic of relations; more exactly, that species of relations known as transformations. This possibility is mooted, though not developed, by Driesch (1908) in his Gifford lectures where he conjoins the idea of a causal theory of morphogenesis, albeit one couched in terms of the concept of entelechy, with the idea of a rational systematics. Thus, whereas the old dialectic was between Darwinian explanatory theory and the Linnaean system, we might envisage a new dialectic between a field theory of forms and a rational systematics. This is what I propose to examine in the next chapter.

5

Rational Systematics and Morphogenetic Theory: A New Dialectic?

Driesch on Rational Systematics

As I noted in Chapter 1, I assume that the question of rational systematics is a particular instance of the more general question of a 'logic of morphology' – the systematic unification of diversity. In the past, and especially in the Continental European tradition, the question of a 'logic of morphology' was regarded as of fundamental importance, not least because it is a presupposition of every theory of descent (see Cassirer, 1950). However, in recent times, at least in the Anglo-Saxon tradition, and consequent upon a preoccupation with the Darwinian Theory of Descent and the belief that the fundamental (or only) relations between forms are those of material continuity, so that diversity can only be unified in terms of historical genesis, that is, genealogically, the question has not been regarded as fundamental. Insofar as it has been considered at all, this has largely been in terms of discussions of the Linnaean hierarchy.

The idea of a 'logic of morphology' seems to have been primarily developed by scholars working within the idealist tradition though some positivistically inclined British philosophers, clearly influenced by the German tradition, have also drawn attention to the problem, in particular Woodger (1945). A more recent discussion by Brady (1987) can also be located within the idealist tradition. Although these discussions are undoubtedly of great interest and importance, they are marred by a tendency to commit the "epistemic fallacy" (Bhaskar, 1978) whereby being and knowledge of being, natural and epistemic possibility and natural and logical necessity are conflated. Now, from a realist perspective in the philosophy of science, the notion of 'logic' pertains only to the epistemic order, the forms of thought in terms of which being is represented or described – the structure of a set of concepts or propositions – and not to the forms of being per se, the ontological order.

It is in this sense of logic that I will discuss the problem. That is, I will be primarily concerned with the kinds of logical structures in terms of which a set, or some aspects of a set, of natural forms, conceived as existing independently of all representation, might be 'rationally' represented as a unified system. However, as will become apparent, the discussion of such logical structures cannot be conducted in complete isolation from questions of ontology and explanatory theory.

As I remarked in Chapter 1, Driesch (1914, p. 140) characterises an empirical taxonomy as a "classificatory *preparation* for the knowledge of . . . the rational in the forms of nature." As I argued, a question of concern is whether or not we can sustain a view of individual organisms as members of natural kinds and, consequently, whether we can regard an empirical classification as providing at least a tentative or approximate representation of a system of natural kinds so that this can provide a starting point for the investigation of any possible rational system. At the same time, we have to acknowledge that the view which conceives such kinds as classes in the sense in which class is understood in traditional formal logic cannot be sustained; a different conception is needed. If a new conception of taxa as natural kinds can be sustained, then the Darwinist position would be inverted; origin or descent would be contingent with respect to identity. The consequence that biology would have two radically different conceptions of taxa is no cause for alarm. Ontologies, that is, natural kind distinctions, are always theory-dependent and, from a realist perspective, there is no general requirement for intertheoretical synonymy (see Bhaskar, 1978)).

As I noted, the question of rational systematics is mooted by Driesch (1908) and characterised as a concern with "diversities." Driesch explains:

All systematics which deserves the predicate "rational" is founded upon a concept or upon a proposition, by the aid of which a totality of specific diversities may be understood. That is to say: every system claiming to be rational gives us a clue by which we are able to apprehend either that there cannot exist more than a certain number of diversities of a certain nature, or that there can be an indefinite number of them which follow a certain law with regard to the character of their differences. (Driesch, 1908, p. 243)

A rational system is thus a closed system. As examples of rational systems or systems which approximate to this ideal in other sciences, Driesch cites theories of solid geometry and of conic sections in mathematics and the theories of crystallography, of the periodic arrangement of the elements and of the homologous series in the physical and chemical sciences.

In Driesch's view, all sciences have their systematic aspect and therefore this would be an important aspect of any would-be science of biological form. Is it possible to construct a rational system of forms? Driesch does not suppose that the domain of biological forms is necessarily rational in its entirety; the project is to discover the extent of the rational, or indeed, if it exists at all. Now, as Driesch (1908, p. 293) observes, from a Darwinist perspective the idea of a rational system of forms is inconceivable. The Darwinist view that forms are accidental implies, firstly, that an indefinite number of forms is possible and, secondly, that there is no law relating these forms; in other words, the world of forms is conceived as an open system. Now, of course, there is, in principle, an indefinite number of molecules which belong to any homologous series, but they all conform to the law of the general molecular formula. Consequently, the significant feature of the Darwinist conception is not the indefinite number of possibilities but the absence of law. It is this, according to Driesch, which results in systematics losing its fundamental importance and becoming a "mere catalogue."

Let us now explore in more detail the nature of rational systems. As Driesch points out, the general logical type of all rational systems (as opposed to merely empirical classifications) can be specified a priori.

Rational systematics is possible whenever there exists any fundamental concept or proposition which carries with it a principle of division. . . . The so-called "genus" . . . then embraces all its "species" in such a manner that all peculiarities of the species are represented already in properties of the genus, only in a more general form, in a form which is still unspecified. The genus is both richer in content [i.e intension] and richer in extent than are the species. (Driesch 1908, p. 245)

This is to be contrasted with the situation which pertains with regard to classifications which are merely empirical, such as the Linnaean hierarchy, where

genera and species. . . . can be related only on the basis of empirical abstraction . . . here, indeed, the genus is richer in extent and poorer in content than are the species. The genus is transformed into the species not by any inherent development of latent properties, but by the mere process of addition of characteristic points. It is impossible to deduce the number or law or specifications of the species from the genus. (Driesch 1908, p. 245)

The logical problem of rational systematics, therefore, concerns the nature of the 'generic concept' and the distinction between the 'rational generic concept' and the 'empirical generic concept'. It would seem that, in his formulation of the problem, Driesch has in mind Hegel's distinction between two forms of the concept, the "Concrete Universal" and the "Abstract Universal" (see Taylor, 1975). For Hegel, the Concrete Universal is "the inner principle of a diversified totality." Here, the concept (for example, the genus) is internally, that is necessarily, related to its species in that the number, differentiae and articulation of the species can be deduced from it. The extension of the concept thus forms an ordered totality or systematic whole; it is itself a particular. In contrast to the Concrete Universal, the Abstract Universal is thought in abstraction from concrete specificity and individuality. The relation between genus and species is external – that is, contingent – hence the species cannot be deduced from the genus, specific differentiation is arbitrary as regards nature and number and species are not necessarily articulated with each other. Rather than an ordered totality we have simple diversity or multiplicity.

In his Gifford lectures, Driesch (1908) is primarily concerned with the causal problem of morphogenesis and his consideration of the possibility of a rational system of forms is relatively brief. In relation to the nature of Abstract and Concrete universals, his discussion amounts to little more than a comparison of the two kinds of concepts and the assertion, as I have noted, that the latter is the logical form of all rational systems.

Cassirer and the Concept of Form

For a more detailed and profound analysis of nature of universal concepts, and in particular the Concrete Universal, we have to turn to Cassirer, whose preoccupation with the problem extended over three decades (Cassirer, 1923, 1944, 1957). Indeed it is one of the central aspects of his own systematic philosophy, and even the necessarily brief outline which I present here should make clear why Cassirer has been called "the founder of philosophical structuralism" (Caws, 1988). A major point of his argument is that the traditional logic of classes is dependent upon or tacitly presupposes the more fundamental logic of relations because every class can be defined as a series whose members are related to each other by a principle or "generating relation" (see also Krois, 1987). In his early work he explored the nature of the concept in the modern mathematical and physical sciences where the relational nature of the universal concept is made explicit and the "exact" scientific concept takes the form of a kind of Concrete Universal. In brief, in the mathematical and physical sciences there is a shift from the 'generic concept' in which the idea of a kind is the idea of a class to the 'serial concept' in which the idea of a kind is the *form* of a series; in effect, a shift from traditional classes to equivalence classes of a particular kind.

At the heart of the systematic problem is the problem of the relation between the particular and the universal as it manifests itself in the work of "reflective judgment," to use the Kantian expression. That is, the form of judgment in which the particular is given for which the universal has to be found. Consequently, as Cassirer (1981) explains, the problem of judgment is joined with the problem of concept formation, for it is the concept that 'groups' particular cases into a higher genus, thinking them as contained under its generality. In his 1923 discussion of concept formation in mathematics and the physical sciences, Cassirer develops a wide-ranging psychological and logical critique of the empiricist doctrine according to which universal concepts are supposed to be formed by the abstraction of what is 'common' from a set of entities taken as given. Some aspects of Cassirer's critique involve somewhat similar points to those made by Driesch. For example, he draws attention to the 'irrationality' of the abstract, generic concept in the sense of the 'gap' between the universal and the particular, the fact that the species cannot be rationally 'recovered' from the genus. In this connection, Cassirer makes the significant observation that the 'irrationality' within the empiricist doctrine of the concept is a consequence of retaining Aristotle's logic but discarding his view of the real structure of nature, his ontology. In the original system, the ontology complements, as it were, the logic, and compensates for the 'irrational' element in it, since it is supposed that there is a natural, teleological movement or development from *genos* to *eidos* so that the 'irrational' element of the logic is, effectively, eliminated or bypassed.

However, Cassirer's critique goes further than that of Driesch. The doctrine of abstraction cannot be regarded as an explanation of the formation of concepts, for in that doctrine the concept, at least in its fundamental *function*, is tacitly presupposed. The doctrine supposes that the common features are to be discovered by means of comparison, which activity is guided by the similarity between things. But 'similarity' is not a characteristic of a thing. The doctrine of abstraction covertly introduces this *relation* between things into the observable qualities of a thing and then claims to rediscover it. Cassirer argues that any comparison of the contents of perception with a view to abstraction presupposes a constructive act of identification in which these contents are thought *from the first* not as disconnected particularities but as an ordered manifold. That is, elements are conceived as being related in some definite way such that they comprise a necessarily connected *series* in which the identity of the relation between the elements is preserved despite changes in the concrete content. Concepts are the rules or laws that generate series, not members of the series. In this sense the concept in its basic function – the unification of a multiplicity – is a presupposition not a consequence of abstraction. In fact Cassirer argues that it is entirely contingent whether an act of abstraction actually occurs; that is, an act which results in the formation of an 'abstract object' in which similar features are united – a generic concept in the traditional sense.

As he observes, a further weakness of the theory of abstraction is its selection of only *one* of the many possible principles of logical order, that is, similarity. However, the conceptual ordering of a series could equally well be accomplished in accordance with, for example, a principle of difference as Lévi-Strauss (1969) has recently reminded us.

The universal concept, in its fundamental nature therefore, is the "form of a series" – that is, the law of relation or connection between the elements which comprise a series and by means of which they acquire their cognitive identity. Such a series is not 'given' but constructed; in the language of phenomenology, it is an intentional object (see also Brady, 1987).

Some two decades later, Cassirer (1957) returns to a consideration of the nature of the concept construed as the form of a series. He sees his earlier arguments as being confirmed by recent developments in logic and now explicitly situates his views in relation to the Kantianism which remains his point of reference. Other logicians, it is claimed, have argued that we should abandon the view in which a concept is necessarily equated with the idea of a class and in which all relations among concepts are conceived as being reducible to those of subsumption or inclusion of species in genera. Rather, we should adopt the much more general, Kantian view of the universal concept "as nothing other than the unity of rule by which a manifold of contents are held together and connected with one another" (p. 287). The formation of the universal concept does not involve the formation of some kind of "generic image" by discarding the particular determinations that are contained in the individual, concrete images; rather it involves positing the particular determinations as variable. Thus the various structures which we regard as examples of one and the same concept are connected by means of a "rule of change on the basis of which one example can be derived from another . . . up to the totality of all possible examples" (p. 291). Consequently, with regard to the relation between unity and multiplicity, 'one in many': "the One is not so much a unity of the genus under which the species and individuals are subsumed as the unity of the relation by which a manifold is determined as inwardly belonging together" (p. 298).

In a still later discussion (Cassirer, 1944; see also 1979), which is concerned with Group Theory in relation to geometrical and perceptual concepts, he is able to define more precisely the notion of the 'rule' which renders geometrical and perceptual concepts universal. A 'rule' is "that *group of transformations* with regard to which the variation of a particular image is considered." (p. 22) As he observes, the idea that a universal geometrical concept, for example that of a triangle, could take the form of a sort of generalised image, generated by abstraction from various images of particular, concrete triangles, presents us with a "problem that is logically impossible and psychologically insoluble. But it is quite a different matter to start from the intuition of a given concrete figure and at the same time conceive in the latter the totality of possible transformations to which it may be subjected according to certain laws of transformation," (p. 23). We can anticipate a little by noting that this is akin to the procedure followed by D'Arcy Thompson in generating his famous biological (shape) transformations. The various figures generated in this way would be equivalent under transformation; that is, they would be members of an equivalence class.

As Cassirer (1944) observes, very different concepts may be generated on the same perceptual base by the application of different groups of transformations, that is, by the insertion of perceptual constants into different structures. Earlier Cassirer (1955) had made the same point, informally, with respect to language, and a broadly similar position characterises Harrison's (1979) view of ordinary language. The latter argues explicitly that linguistic discrimination cannot be identified with perceptual discrimination for perceptual similarity is ubiquitous; any two entities regarded from some point of view will resemble each other in some respect. The determination of identity, therefore, implies a *specific* point of view, an "identity of reference" (Cassirer, 1923), and such a point of view *is*, functionally, a concept.

In short, kinds are not given but theoretically constructed. As Cassirer (1944) observes in relation to geometry, Group Theory enables us to give a precise specification of the 'point of view', the system of reference, in terms of which different particular figures may be compared, hence kinds of figures constructed. He summarises thus: "the real foundation of mathematical certainty lies no longer in the elements from which mathematics starts but in the *rule* by which the elements are related to each other and reduced to a 'unity of thought' " (p. 8). The concept of transformation thus provides a means for the systematic unification of diversity and assertions of 'sameness' and 'difference' are dependent upon which 'rule', that is, group of transformations, is chosen. Thus from the perspective of Metrical (Euclidean) geometry, the different conic sections are indeed distinct entities, different in kind. When we move to Affine Geometry, however, the (apparent) "essential" difference between the circle and the ellipse, disappears; they are now "essentially" the same. This development is carried further in Projective Geometry, where *all* the conic sections are regarded as the same kind of figure. Moreover, as Klein's 'Erlanger Programme' demonstrated, the different geometries comprise a hierarchy and themselves can be related as one integrated system of transformations (see also Piaget, 1971).

As Cassirer (1923) points out, the concepts of pure mathematics are not and cannot be formed by abstraction, because there is no concrete correlative in the 'given' facts of nature to such concepts as a dimensionless point or a perfectly straight line. Rather they are theoretical constructions, and, argues Cassirer, the same is true of the concepts of the physical sciences.

The mathematical concept as embodied in a formula or equation seems to be a kind of Concrete Universal. As Lambert pointed out in the eighteenth century, the peculiar merit of the 'general' mathematical concept, unlike the abstract concept, is that it does not eliminate the determinations of the special cases; rather they are retained and, moreover, are deducible from the general formula. The mathematical concept, therefore, is not a generic concept in the traditional sense (that is, an Abstract Universal) and the relation between the formula and the special cases is not one of inclusion or subsumption. Rather, a mathematical concept such as the general equation of the second degree with two unknowns (the general quadratic equation) is a kind of Concrete Universal; a concept which contains within itself all the specific determinations for which the formula in question holds. It is universal in the sense that the formula is a symbolic representation of the law or principle of order which determines and relates *all* the conic sections; that is, it is a representation of the form of a series of geometrical figures. It is concrete because when the variables are given particular numerical values, the different conics are generated as species.

For Cassirer (1923), the mathematical concept represents the ideal of the scientific concept, which he opposes to the generic concept – the schematic representation of the abstracted 'common element' in a series of similar particular things. The 'exact' scientific concept pursues a similar intellectual goal to that which exists in pure mathematical knowledge in that the scientific concept does not eliminate or disregard the particularities which it represents. Rather, it shows the *necessity* of the occurrence and mode of connection of just *these* particularities; "what it gives is a universal *rule* for the connection of the particulars themselves ... the individual case is not excluded from consideration, but is fixed and retained as a perfectly determinate step in a general process of change" (p. 20), or more generally, variation. Thus, although the characteristic feature of the theoretical, scientific concept is "the universal validity of a principle of serial order" (p. 20), the emphasis, as compared with the generic concept, has shifted from universality to necessity; that is, the necessity of the serial order of the particulars which are its mem-

bers. Since this necessity pertains to relations within the domain of representation it is a species of logical, not natural (causal), necessity.

The point can be illustrated by reference to the transformation of chemistry
from a purely empirical science to a rational, theoretical science which is
effected, as Cassirer (1923) argues, by means of serial concepts. A good
example (though not one discussed in detail by Cassirer) is the one cited as
a paradigm of a rational system by Driesch, namely the homologous series.

An homologous series is a series of compounds of uniform chemical type
which show a gradation in physical properties (e.g., boiling point). An example is the alkanes (paraffins), where the series includes methane, ethane,
propane, butane and so on. The alkane series can be represented by a general
molecular formula which shows the kinds and numbers (but not the arrangement) of the atoms present in the molecule: C_nH_{2n+2}. Thus, to emphasise
what is at issue (and at the risk of doing some violence to the history of
chemistry), we can regard 'paraffin' as a rather crude empirical, generic concept which subsumes substances with roughly similar chemical properties.
This is replaced by the rational, scientific concept of the alkanes, expressed
in terms of the general molecular formula, which is not a generic concept
but a serial concept. The exact scientific concept is a symbolic representation
of the form of the series. In this sense the formula is an "analogue of mathematics" (Cassirer, 1923, p. 219) for by substitution of numerical values for
'n' any member of the series can be generated. The series as a whole thus
constitutes a closed set (though not, I think, a group) with an indefinite number of members (species) and the members are related in terms of transformation. Thus the form of the series is constitutive of the alkanes as an object
of theoretical knowledge; conversely, the distinctive identity of the alkanes
as a kind lies in the form of the series. Other series have different forms,
hence are different kinds: the alkenes (olefins) have the serial form C_nH_{2n}
and the alkynes (acetylenes) the form C_nH_{2n-2}. Thus the general molecular
formula functions as the equivalent of a definition of the kind since it is a
symbolic representation of the invariant which characterises the kind; in this
sense it could be regarded as the "essence" of the kind. Hence, while there
are 'limits' to variation, insofar as any entity can be represented in terms of
this invariant, then that entity is a '*typical*' member of the kind. In relation
to this, a significant property of the 'serial form' is that *any* member of the
series implies the other members (as possibilities) and, in so doing, can function as a *concrete symbol* of the kind, that is, can represent the kind (see
Cassirer, 1923). The possible significance of this point in relation to the role
of the holotype individual in empirical taxonomy is evident.

Goethe and the Phenomenology of Form

It is one thing to explicate the logical form of rational systematics; it is quite
another to determine whether the notion has any legitimate or significant
application to the world of organic forms. To pursue this question, the scope
of the discussion must necessarily broaden, for considerations in addition to
those of logic are relevant. Cassirer (1923, p. 220) notes that the transformation of empirical diversities into rational, intelligible diversities does not
bring a science to a conclusion. Subsequent to this, in his view, there is the
problem of understanding and grounding the "laws of structural relations"
in terms of "deeper" causal processes. Since from my perspective, description and explanation are related dialectically and not successively, a consideration of any possible rational systematics must, from the beginning, take
into account the question of explanatory theory.

The view that the diversity of organic form might be rendered intelligible
in terms of the concept of transformations is, of course, not a new one in
biology. The important concept of the 'serial form' seems to have been anticipated, if only intuitively, informally and obscurely by Goethe in his Theory of Metamorphosis (Goethe, trans. Arber, 1946; see also Giacomoni, 1993)
which was devised, at least in part, in response to the difficulties of identifying and classifying plants in terms of the Linnaean system (see Cassirer,
1950). For Cassirer, the distinctive feature of Goethe's Concept of Form is
the relation it posits between the "particular" and the "universal." He argues
that Goethe conceives the particular and the universal as not simply connected; rather "they interpenetrate one another. . . . The "factual" and the
'theoretical' [are] not opposite poles . . . but only two expressions and factors
of a unified and irreducible relation." As Goethe puts it: "The particular
always underlies the universal; the universal must submit to the particular."
For Goethe, as interpreted by Cassirer, the relation between the two is not
one of logical subsumption but of ideal or "symbolic" representation. As
Goethe explains: the particular represents the universal as "a momentarily
living manifestation of the inscrutable"; it exhibits the "immanent law" of
nature. Or, "the spirit of the actual is the true ideal." The universal or generic
concept of 'type' must be conceived, in Goethe's words, as "a real Proteus,"
a view which seems paradoxical from the traditional (Cuvierian) perspective
in which the being of the 'type' is understood as the being of a thing – a
'static', abstract schema. For Goethe, however, the being of the 'type' is
conceived as the being of a synthesis. On the basis of some general remarks
of Cassirer concerning representation (1957, pp. 200–202), it is arguable, that,
for Goethe, this synthesis is, or is initially, a perceptual-aesthetic synthesis

in which a set of forms, for example, the sequence of leaves along the axis of a plant, is grasped intuitively as an 'animated totality'. The plant 'lives' in me as it were, an idea which is not quite as bizarre as it may first appear. However, with a shift of perspective to another modality of meaning, the set becomes intelligible in logical and conceptual terms as a concrete, 'dynamic' schema which represents a law. Consequently, for Goethe, the task of the morphologist is to grasp, intuitively and/or conceptually, the 'unchanging' – the "spirit" or the "immanent law" – which reveals itself in change; in our terms to grasp the 'form of the series.' Once in possession of this knowledge, as Goethe (1970) observes in a letter to Herder, "It will be possible to go on for ever inventing plants and know that their existence is logical; that is to say, if they do not actually exist, they could, for they. . . . possess an inner necessity and truth" (p. 310). Thus, to employ the terms of the present discussion, Goethe appears to envisage, albeit somewhat obscurely, the construction of a rational system in terms of which diverse forms can be derived as possibilities.

As regards Goethe's theory of metamorphosis, although this represents particulars as genetically connected, the "genesis" here cannot be understood as real and historical. The "genetic" transformation in terms of which particular organs of a plant are supposed to be related to each other is an ideal, not a real genesis and the relation takes the form of a "law" which appears to define a kind. The concept of a 'type' or kind is, therefore, the concept of a "law" in terms of which all the variants of that kind can be intelligibly connected; they are all 'typical forms'. Given that this conception seems to have been forgotten in the mainstream of biology (if it ever entered), it is significant that Darwin (1859, p. 417), who, following Cuvier and Owen, conceives the notion of 'type' in terms of a static 'schema', evidently misunderstands Goethe's theory of metamorphosis – Goethe is not named, but the reference is unmistakable – and supposes him to be talking about real and historical rather than ideal, that is, conceptual connections between forms.

Goethe's morphological concepts have recently been analysed and contrasted with those of Owen in an important and illuminating paper by Brady (1987). In his reconstruction, Goethe's morphology is represented as a phenomenology of form in which the unification of a set of diverse forms is achieved at the level of perception by intending (in the phenomenological sense) the set as a transformation series; that is a set within which we experience a "movement" or transition from one member to the next (compare the notion of an "animated totality" mentioned earlier). As Brady observes, such a concept of unification is quite different in kind from that of Owen and Darwin, which is based on the notion of "theme and variations," where the "theme" is conceived as an abstract 'schema'. Brady's reading appears to be supported by Goethe's own remarks and Cassirer's analysis and, insofar as I am competent to judge, is convincing

Brady characterises Goethe's morphology as a descriptive science which does not attempt "explanation" insofar as this involves the construction of an additional explanatory theory by means of a "speculative act." However, he also claims that this kind of phenomenology has "causal implications" in that the description is of such a kind that it is also explanatory. His argument is somewhat terse, but, insofar as I can follow it, it seems to hinge on the existence of the relation of necessity within the serial form. I have claimed that this necessity is logical, not natural or causal. Brady seems to suggest that on the Goethean view, logical and causal necessity are indistinguishable, hence the serial form is at one and the same time both descriptive and explanatory. This claim involves such a radical departure from most conventional views, which emphasise the sharp distinction between logical and natural necessity, that, as Brady is aware, it requires a far more extended discussion than he provides. On the basis of the discussion he does provide his conclusion appears unwarranted, or at best premature. Whereas the relation of logical necessity permits thought to 'move' through a series in either direction ($A{\rightarrow}B$ or $B{\rightarrow}A$) the relation of causal necessity is characterised by unidirectional succession ($A{\rightarrow}B$). Brady restricts himself to a consideration of what he calls a "time form," that is a transformation series in which the members actually develop successively so that the series manifests itself progressively in time; the example he uses is one of Goethe's own, the proximodistal appearance of leaves on a single plant. Here we seem to have a combination of a unidirectional succession in time with a relation of necessity, hence, apparently a causal relation. Thus, as I read him, Brady seems to claim that, purely by means of experience, we can intuitively grasp the actual productive or morphogenetic process occurring in the plant together with the natural possibilities inherent in this process. However, if I have understood it correctly, it would appear that the argument is flawed, both in the specific case considered by Brady and more generally and seems to involve the "epistemic fallacy" characteristic of the idealist tradition. In the first place, one mature leaf does not actually produce, or change into, the next in the series; the 'generative relation' or 'dynamic' is purely epistemic, as are the possibilities; both exist entirely within the domain of representation. In the second place, although it is true that the leaves are sequentially produced in time by the plant, temporal order plays no significant or essential part in our experience of the set of leaves as a transformation series since, according to Brady's own account, this experience is of the complete (or almost complete) set of leaves arranged in space on a plane surface. Hence, 'movement' can be experienced in either direction through the series. Moreover, even if the argument were tenable in the case considered by Brady, it is not generalisable for it is evident from the biological transformations considered by D'Arcy Wentworth Thompson (see the next section) and from Cassirer's discussions of a wide range of phenomena that the concept of the serial form is not restricted to series whose members come into being successively; consequently, in these other series, temporal order is completely irrelevant. If this is so, then Brady's claim, construed as a general claim regarding the explanatory nature of this kind of phenomenology, is unjustified and we are left with a 'pure' description, that is a classification of forms, albeit a classification of a rather distinctive kind.

Now, I have no wish to criticise a descriptive phenomenology of form per se. Indeed, it is arguable that our capacity to intend (in the phenomenological sense) a 'unity within diversity' plays a crucially important role in perceptual recognition and in certain forms of aesthetic experience (see Cassirer, 1944; also Lévi-Strauss, 1966, 1969; for a critical perspective see Hagen, 1980, 1986). Moreover, it is conceivable that such phenomenological understanding may be an important component of the kind of tacit, "practical-aesthetic" knowledge ("knowing one's organisms") which is characteristic of biological practice in both Natural History and experimentation, but which rarely finds explicit or formal expression on the printed page (see also Grene, 1974). However, the crucial question is whether morphology, construed as a *science* of form, can remain content with the phenomenological description and unification of experience and eschew "speculative" theory. I argued in Chapter 4 that it cannot and that "speculative," explanatory theory is indispensable to science. The phenomenological approach to form seems to be an attempt to construct a natural order of forms entirely within the domain of material practice; that is, a practice concerned exclusively with those properties and dispositions of entities which are available to experience. In Brady's phenomenology, immediate experience untainted by "speculation" seems to be the analogue of the "neutral" observation language beloved of positivist philosophers of science. Such a phenomenology is, therefore, an example of positivist "Machismo," to employ Ernest Gellner's witty term; the attempt to build a science on the (supposedly) *certain* foundations of experience. However, part of the thrust of the realist critique of positivism (see Harré, 1970, 1986; Bhaskar, 1978) is that there are no such incorrigible foundations, for the scientific enterprise is both inherently "speculative" and inherently historical. The progress of a science involves a dialectic between taxonomic (descriptive) and explanatory knowledge and knowledge of both kinds is defeasible. Consequently, the development of a science may require that violence be committed on the apparent truths of experience in an effort to grasp the real truths of the world.

As I noted earlier, from a realist perspective, the manifest properties of entities which are available to experience in material practice have to be understood as the realisation of dispositions which are grounded in the natures of things. Natures are determined by structures which are 'hidden' and, as such, not (immediately) accessible to experience; they have to be constructed by the speculative work of theoretical imagination. The goal of the systematic enterprise of science is the construction of a real order of natural kinds or taxa. As Harré (1986) argues: ''If taxa are to be reliable they must mark off natural kinds. But natural kinds can be demarcated in a stable way only if each natural kind concept can be located in both a practical and a theoretical context.'' In other words, the construction and demarcation of real kinds cannot be accomplished without reference to causes; that is, to the properties and structures of things which are beyond actual experience and which are supposed, theoretically, to account for the manifest, experienced properties. The inadequacies of phenomenology are revealed by a consideration of the systematisation of the chemical elements in the periodic table where the discrete classification is grounded in the electronic theory of the chemical atom so that each element in the series differs from the next by just one electron. In the absence of theoretical considerations, classification of the elements would be, to some degree, indeterminate as I noted previously.

A Goethean phenomenology of forms, as reconstructed by Brady, raises somewhat similar problems. For instance the criterion for inclusion of an entity within a 'transformation series' is purely formal; the entity must conform. That is, inclusion should strictly be determined only by the 'form of the series.' In the example discussed by Brady, the series is one of perceptible shape transformations; consequently, any entity of whatsoever nature could, and should, be included provided only that it obeys the law of the series. In point of fact, Brady includes only leaf forms – biological entities – in a particular series; he excludes artificial forms (replicas of leaves, say) and crystalline forms (some of which resemble his leaf forms). In other words, he tacitly invokes a criterion for inclusion and exclusion other than, and in addition to, that provided by the purely descriptive phenomenology of form. In this case, it would appear that the criterion is that the entities must be biological entities, that is, entities in which the causal mechanisms responsible for the production or generation of the manifest forms might be assumed to be different in nature from those involved in the production of the forms of artificial or inorganic entities. In like fashion, albeit somewhat inconsistently, D'Arcy Thompson employs tacit, nonformal criteria for exclusion of entities from his transformation series.

The concept of transformation, as conceived from this perspective of 'rational empiricism', raises the general problem that if we confine ourselves to the realm of experience, (almost) anything might be represented as possessing some formal relation to (almost) anything else when regarded from some point of view as Cassirer, in effect, observed. But can we conclude from the existence of such a relation that we are dealing with a single natural kind? It is by no means clear that we can draw this conclusion, anymore than we can conclude that the existence of general term in ordinary language necessarily implies the existence of a single natural kind. Thus, the terms "Dog's Tooth Violet" and "Sweet Violet" do not denote variants of a single natural kind in any biologically significant sense (see also Dupré, 1981). As Bhaskar (1978) observes, not all general features of the world have a common causal explanation.

It would seem that difficulties of the kind outlined can only be avoided if we adopt a determinate point of view, an "identity of reference" in Cassirer's terminology, and I suggest this can only be accomplished in a nonarbitrary way in biology as it has been accomplished in the physico-chemical sciences, that is, by explicitly (rather than tacitly) invoking a "speculative," causal (generative) theory which makes reference to natures. Such theory cannot be regarded as simply subsequent or "supplementary" to a description regarded as completely achieved. Rather, the questions of logic and systematic representation which pertain to description must, in principle, be conceived in *dialectical* relation with the questions of ontology and the consideration of natures and causes which pertain to explanatory theory. A "theoretical critique of the facts" (Mepham, 1972) is as necessary and legitimate as a factual critique of theories.

Similar conclusions result from a consideration of the limitations of any kind of phenomenological, purely descriptive, analysis of form. Goethean phenomenology seems applicable only to continuously varying forms; that is a series of forms must be such that a continuous, uninterrupted "movement" can be experienced (see Brady, 1987; compare D'Arcy Thompson, 1942, p. 1035). As such it seems restricted to what Bateson (1894) calls "substantive variation," that is, variation in the properties of morphological elements such as shape, size, colour and so forth. Yet much morphological variation – Bateson's "meristic variation" – is discontinuous. For example, variation in the number of parts, such as the number of vertebrae or segments in an animal or the number of petals in the flower of a plant. It would seem that a unification of diversity of this kind cannot be achieved by any means which restricts itself to the level of experience. We cannot, for example, generate the diverse forms by means of a simple arithmetic rule involving the operation of addition, subtraction or multiplication of perceptible morphological elements as Darwin (1859, p. 417) seems to suppose. For to do this presupposes the existence of an abstract 'schema' such that the individual elements in the diverse forms can be identified with each other, hence the 'additional' (or 'missing') elements distinguished from the 'original' (or 'remaining') elements, using purely empirical criteria. But in many cases of meristic series, including those cited above, such identification is impossible, as Bateson (1894) points out (see Chapter 2). He argues that such discontinuously variable forms must be conceived as wholes, that is, systems of relations or patterns, not as variable aggregates of independent, individual elements. Conceived in this way, the change from, say, five to six petals in a flower is a change from fivefold to sixfold rotational symmetry and this involves the *replacement* of one kind of thing (a pattern) by a completely new kind of thing, a mode of change which is unintelligible. Bateson suggests that such forms can be intelligibly related only by means of "speculative" theories concerning the causal or generative mechanism which produces the empirical forms and in terms of which they can be seen as alternative actualisations of natural possibilities; alternative "Positions of Organic Stability" (see Webster, 1992). In other words, the discontinous and unintelligible empirical change must be conceived as the 'expression' of an intelligible change in a 'hidden' thing; that is, a change in which some element of continuity is preserved such that the thing can be conceived as remaining fundamentally the same kind of thing. As I have argued, in this book the 'hidden' thing is conceived as a morphogenetic field. I should, perhaps, add that Bateson's denial of the independence of morphological elements pertains to them only insofar as they are considered as elements of a meristic series. He does not wish to deny, indeed is concerned to emphasise, that in other respects, viz with regard to "substantive" (qualitative) variation, they can sometimes behave independently.

It seems to follow from the argument presented in this section that, if the notion of transformation is to be retained, we have to shift our attention from a consideration of empirical forms as transformations to a consideration of the generative structures – that is, morphogenetic fields – as transformations. Any rational system must be constructed at a 'deeper' level of reality.

On Biological Transformations: D'Arcy Thompson

The most famous exposition of the 'transformational' view in modern times is, of course, that of D'Arcy Thompson (1942) behind whose "rational empiricism'' it is not difficult to see the shadow of Goethe. Indeed, D'Arcy Thompson explicitly denotes (p. 1026) his own study of organic form by "Goethe's name of Morphology" while at the same time explaining that this is only a part of "that wider Science of Form which deals with the forms assumed by matter under all aspects and conditions.'' It is no part of the object of this book to expound or defend a general science of form. There can be little doubt, however, that D'Arcy Thompson's commitment to this wider enterprise accounts, at least in part, for the strongly reductive tendencies in his thought.

In the present context, D'Arcy Thompson's concern with what, quoting Bergson, he calls the "logical affiliation between forms" and his aim to demonstrate that "natural differences might be reduced to rules" (p. 1055) explicitly suggests a means of moving away from the Darwinist preoccupation with kin and contingency and a return to the pre-Darwinist concern with kind and necessity. As such, the importance of his work can hardly be overestimated, despite its preliminary and programmatic nature and even if his own employment and, indeed, conception, of the 'transformational method' is fraught with difficulties and limitations, many of which are consequent on its positivist nature.

Indeed, positivist "Machismo" is a strong motivating force behind D'Arcy Thompson's antipathy to Darwinism and comes across clearly in his remarks on the Foraminifera:

While we can trace in the most complete . . . manner the passage of one form into another . . . the question stares us in the face whether this be an 'evolution' which we have any right to correlate with historic time. The mathematician can trace one conic section into another and 'evolve' for example, through innumerable graded ellipses, the circle from the straight line: which tracing of continuous steps is a true 'evolution' though time has no part therein. It was after this fashion that Hegel and . . . Aristotle himself was an evolutionist – to whom evolution was a mental concept, involving order and continuity in thought but not an actual sequence of events in time. Such a conception of evolution is not easy for the modern biologist to grasp and it is harder still to appreciate. (p. 869)

While we can admire the clarity with which the nature of a 'rational system' of forms is explicated, it is evident from the context of this passage that D'Arcy Thompson is as much concerned to contrast, in a positivistic manner, the certainties of a mathematical description of what is available to experience with the speculative nature of Darwinian phylogenetic reconstructions which purport to have explanatory significance. Since, as I have emphasised, *all* explanatory scientific theory is the product of the speculative imagination, this characteristic per se provides no grounds for a critique of Darwinism. Moreover, the construction of a merely formal system divorced from any explanatory theory is, as I have noted, fraught with problems. In point of fact, when it suits him, D'Arcy Thompson is as ready as the Darwinists to engage in his own speculations with regard to explanation.

Insofar as D'Arcy Thompson's views can be regarded as merely an explicit and more systematic version of those of Goethe, they are susceptible, as I suggested, to the same kinds of criticisms regarding criteria for inclusion of entities in a particular transformation series. Moreover, D'Arcy Thompson is notorious for his reductive tendency to identify organic and inorganic entities, or at least relatively simple ones, purely on the basis of a formal resemblance which is available to experience; as Russell somewhere remarks, mathematicians do not care what they are talking about. On this basis, he often proposes common causal explanations, thereby implying that such entities are to be regarded as members of a single natural kind. While a few of these identifications and speculative explanations may be correct, others are dubious and some are almost certainly false. For example his comparison of the forms of cells with Plateau's surfaces of revolution suggests to him a common explanation in terms of surface tension. However, measurements of the surface tension of cells indicate values which are too low to have the required effect.

D'Arcy Thompson's most systematic discussion of transformation comprises the last chapter of his book: "On the Theory of Transformations, or the Comparison of Related Forms." Here he considers biological forms, actually shapes, as topological transformations which can be represented in terms of a series of deformations of a system of Cartesian coordinates. While the transformations cannot be conceptually generated since the forms remain undefined and no 'rule' is formulated, the forms can, nevertheless, be compared and the intelligible relations between forms can be 'seen' by means of a series of pictorial representations. What is offered, therefore, is a systematic method of comparing forms. The basic method is, of course, not new. It was employed in a rudimentary fashion by Dürer in the sixteenth century and, in a more sophisticated manner, by Peter Camper in the eighteenth (see Giacomoni, 1993).

The nature and limitations of the 'method of transformations' are analysed in some detail by Woodger (1945). D'Arcy Thompson himself, employing Aristotle's terminology, had drawn attention to the fact that the method is only applicable to "related forms" in which "the parts are identical" and which differ in the way of "excess and defect." In other words, the method can only be employed to analyse "specific" differences within a "genus" in Aristotle's sense of these terms (p. 1034). Woodger explains that the notion of transformation as employed by D'Arcy Thompson involves the idea of one-to-one continuous morphological correspondences between infinite sets of points. Thus he can deal with variations in the *shape* (a nonrelational property) of parts which are in morphological correspondence, that is, parts which are homologous, to use the classical (pre-Darwinian) terminology. The morphological correspondences are the invariants of D'Arcy Thompson's transformations. Thus "related forms" means forms between which there is complete morphological correspondence between the parts ("identity") and which differ ("specifically") in terms of the nonrelational property of shape.

The 'transformational method' as employed by D'Arcy Thompson is, therefore, restricted to taxonomic levels where there is an empirical invariance, a common formal schema, or Bauplan. Where no empirical invariance or common schema exist, as, for example, between members of different phyla, the method of transformations is inapplicable and, as D'Arcy Thompson himself points out, these forms cannot be related as transformations. Indeed, he concludes that such forms are "unrelated things and have come into existence independently of one another" (p. 1086). As he explains:

A principle of discontinuity . . . is inherent in all our classifications . . . nature proceeds *from one type to another* among organic as well as inorganic forms; and these types vary according to their own parameters, and are defined by physico-mathematical conditions of possibility. In natural history Cuvier's "types" may not be perfectly chosen nor numerous enough, but *types* they are; and to seek for stepping-stones across the gaps between is to seek in vain, for ever. (p. 1094)

As I have noted, the 'method of transformations' is a comparative method and the result of its application to empirical forms is an atemporal, that is an ideal, system of forms. Insofar as superficially diverse forms *can* be systematised in this way, that is, referred to identical functions of different co-ordinate systems, then, this constitutes as, D'Arcy Thompson argues, prima facie evidence that "variation has proceeded on definite and orderly lines, that a comprehensive 'law of growth' has pervaded the whole structure in its integrity" (p. 1037). As Needham (1942, p. 128) observes, the holistic character of the organism which is emphasised here suggests some relationship between the concept of transformation – a concept pertaining to description – and the field concept – a concept pertaining to explanation.

Impatient for explanation, and in characteristic fashion, D'Arcy Thompson leaps from the construction of an ideal system of forms to a consideration of the causal explanation, in "physical" terms, of *actual* change from one form to another. Thus he claims (p. 1033) that once a set of forms can be shown to be transformed representations of each other "it will be a comparatively easy task (in all probability) to postulate the direction and magnitude of the force capable of effecting the required transformation.''

However, as Woodger observes, the forms analysed by D'Arcy Thompson are (for the most part) 'adult' forms, what Woodger calls "taxonomic transformations," and no one supposes that one such form actually and directly changes into another. Consequently, Woodger argues that the method of transformations needs to be developed so that "taxonomic transformation" can be understood in terms of other transformations. According to Woodger, these are of two kinds. Firstly, *embryological* transformations which characterise relations between 'time slices' of *single* individual lives (*A, B, C,* etc.) and which, at least during the earlier stages of development, would have to be of quite a different kind from D'Arcy Thompson's shape transformations. Secondly, *genetic* transformations which characterise relations between 'time slices' of different zygotes (*A', A'', A''',* etc.); 'genetic' here is used in a literal sense rather than in the restricted, technical sense pertaining to the theory of the gene (see subsequent discussion). Woodger argues that "taxonomic transformations" can be formally expressed as the relative product of these two kinds of transformations. Thus, the explanation of taxonomic differences depends upon showing how one zygote must differ from another zygote in order that an adult 'time stretch' reached from the one will differ from an adult 'time stretch' reached from another.

Woodger argues that this procedure could, at least in principle, overcome some of the limitations of D'Arcy Thompson's approach in that it might be possible to relate those forms which, as *adult* forms, appear *unrelated,* and thus eliminate the 'discontinuity' which D'Arcy Thompson concluded must exist between different 'types'. For it might be the case that the structures characterising the zygotes of such differing adult forms as the cuttlefish and the beetle could be related as transformations even though the adults could not.

Woodger summarises his view of the goal of a 'science of transformations':

This points the way to the future structure of biological theory. We require a means of analysing and expressing embryological transformations so that given a certain sort of zygote and a certain sort of environment we can predict the sort of time-slice that will be realised in a given time. We also require a corresponding theory of genetic transformations. When biological theory reaches this advanced stage all the taxonomic transformations will be immediately deducible from the embryological and genetic ones. . . . we shall be able to replace the classification of adults by the classification of zygote types, and phylogenetic speculation will be transferred from the plane of taxonomic transformations between adults to the plane of actual or possible genetic transformations between zygotes. (pp. 117, 120)

A Realist Science of Transformations

It is apparent from the preceding quotation that Woodger characterises the goal of a 'science of transformations' in positivist terms as involving the construction of a formal system by means of which diverse particular forms can be derived as possibilities of that system; this is, in effect, the realisation of Goethe's project as outlined in his letter to Herder and noted earlier. However, the 'realist element' in Woodger's thinking consists in the fact that he conceives this construction as being *dependent* upon a theory of morphogenesis which refers to 'hidden' structures. In this connection he argues that the characterisation of embryological and genetic transformations requires a "theory of zygote structure" and that to achieve this, the existing "theory of the gene" must be extended to include a "theory of cytoplasmic organisation." This view can, with advantage, be modified and generalised for not all new individuals come into existence by means of zygotes and, in any case, as Woodger points out, the zygote *is* the organism (p. 95). Consequently, we might rephrase Woodger's demand as a requirement for a theory of the morphogenetically relevant structures of the organism. These structures must be conceived as dynamic since, as well as being reproduced during the normal course of development whereby a part becomes a manifold whole, they must also be reproduced during the regulatory events that follow experimental perturbation. As I have stated, the position adopted in this book is that the causal mechanism responsible for the production of empirical morphologies is that structure of the organism referred to in the classical literature as the morphogenetic field. I have also suggested that morphogenetic fields should be regarded as putative natural kinds. From this perspective, the theory required by Woodger is a Theory of Field Structure and, as Goodwin will explain, the embryological and genetic transformations invoked by Woodger should be understood as field transformations.

In the theoretical models proposed by Goodwin, the field is conceived as a dynamical system and he argues that genetic and environmental factors determine parametric values in the equations describing the field and therefore act to select or stabilise one manifest form from the set of forms which are possible for that type of field. In principle, therefore, a set of determinate empirical forms is conceived as being generated by a series of determinate fields and these fields are the same in the sense that they are all susceptible to description in terms of a single set of fundamental equations. In that sense they constitute a single kind. Consequently, the determinate members of the series are variants of that kind related as transformations – they comprise a rational system – and the basic field equations are, in Cassirer's terminology, a symbolic representation of the form of the series – in a sense, the 'essence' of the kind. Thus, insofar as any set of specifically different empirical forms or patterns can be understood as being produced by the same generative mechanism in this sense, that set can be conceived as comprising a single kind. From this perspective, kinds are theoretically significant kinds, that is, natural kinds.

I have suggested that, despite all the difficulties traditionally associated with it, the Linnaean hierarchy should be regarded as representing some aspects of a real natural order, albeit approximately and schematically. In terms of our starting point with the problems posed by Driesch, the question therefore arises as to the relation between this purely empirical classification and a possible rational systematics of the type outlined.

Empirical taxonomies attempt to systematically classify organisms in terms of some form of 'sameness' which ranges from that which is general and widely shared to that which is specific and shared by relatively few individuals. In 1828 von Baer proposed, on theoretical and empirical grounds, that the morphogenetic process in an individual organism proceeds temporally from the general to the specific. That is, the basic body plans and axes of symmetry which are the empirical characteristics of the higher taxonomic group to which the organism belongs seem to be established first whereas the specific characteristics and individual features are produced later (see Driesch, 1908; Russell, 1916; Lovejoy, 1968; Løvtrup, 1984). Thus, it is claimed that, in the temporal course of morphogenesis, the individual organism effectively 'runs through' the hierarchy of taxa; Løvtrup makes essentially the same claim. This claim must be sharply distinguished from the view that 'ontogeny recapitulates phylogeny.' In its received form, von Baer's thesis is by no means unproblematic insofar as it is partly based on what would appear to be, if I understand it correctly, a dubious theory of the 'control' of morphogenesis by "Ideas" (see Russell, 1916). However, a somewhat qualified version of the thesis is suggested by Woodger (1945). According to him, embryonic development proceeds by the "production of parts in existing parts followed or accompanied by the appearance of differences between those parts" (p. 100). Thus, to employ his terminology, 'low level' parts (for example, the brain) appear before 'high level' parts (for example, the eyes) and Woodger argues that 'low level' parts and 'high level' parts belong to different Bauplans which determine different taxa and relate them by inclusion.

Thus, in terms of the type of morphogenetic theory proposed in this book, a qualified version of von Baer's thesis would lead us to suppose the existence of a temporal sequence of kinds of morphogenetic fields where each kind of field corresponds to a level in the empirical taxonomic hierarchy. The evidence from classical experimental morphology is at least consistent with this view (see Chapter 4). Kinds of fields appear to come into being successively and the determination of the developmental fate of parts of the embryo occurs progressively; for example, as noted earlier, it seems that developing material is determined 'generically' as fore or hind limb before it is determined 'specifically' as proximal or distal limb.

In terms of the position I have outlined, a rational system of forms would be a system of fields. Hence, the research project of rational systematics would consist, *in principle*, of an attempt to reconstruct the hierarchical system of empirical taxa as a rational system in terms of the serial forms which are possible for each kind of field in the temporal sequence. It is, of course, impossible to anticipate to what extent, if any, such a project will be successful in practice; if we could do so then, as Driesch (1908) remarks, we would already have what we desire. As he also observes (p. 296), there is no reason to suppose that any system will be rational in its entirety; the project is to discover the 'core' of rationality, if it exists, in the totality of diversities, some of which may be accidental.

Evidently, this is a long-term project and initially we can expect little more than attempts at partial reconstructions of the empirical system; that is, the reconstruction of particular genus–species 'modules' probably chosen from the higher levels of the empirical hierarchy. Goodwin discusses such attempts below.

If this project were to be successful in demonstrating that different fields could be conceived as different natural kinds and each kind corresponds to a part of the empirical taxonomic system then something like the traditional view would be vindicated. However, we should bear in mind that, given the dialectical relation between taxonomic and explanatory knowledge, one outcome of such a project, supposing it to be successful, might be that the merely empirical classification would be subject to theoretical corrections of a more-or-less radical nature, as has happened in the physical sciences, since the criterial ('essential') properties used for distinguishing kinds will be those which are grounded by the explanatory theory. Thus, what from an empirical perspective may appear to be two distinct kinds may come to be regarded, from a theoretical perspective, as simply variants of a single kind and vice versa.

A 'rational' perspective would enable us to clarify some aspects of taxonomic practice in that it permits a clarification of the notion of a 'normal' form and enables us to distinguish it from the idea of a 'typical' form. On this view, a taxon is a system of transformations; that is, it corresponds to a series of fields which are equivalent under transformation. As I have emphasised, in such a series all members are equally 'typical'. Thus there can be no such thing as an 'abnormal' member except in the sense of a form which is statistically infrequent. Now, Goodwin will argue in Part II that morphogenetic fields are kinds of dynamical systems which exhibit what he calls "robust" properties. What this means is that, although variations of manifest, empirical form are a consequence of parametric variation, the same manifest form can be produced over a range of parameter variation. Thus one might expect some empirical forms to be statistically frequent, hence 'normal' whilst others are infrequent. Consequently, while the latter individuals are 'abnormal', they are, nevertheless, as 'typical' as any other individuals. Hence they could, for example, serve as holotype specimens in taxonomic practice.

It is apparent from the quotation from Woodger given previously that he conceives a science of transformations as being pertinent to Darwinian theory and "phylogenetic speculation," and this is no doubt correct. However, while the reconstruction of phylogenies is of fundamental importance from a Darwinian perspective, since forms can only be unified in terms of their historical genesis, this is not the case from the perspective of rational systematics. It is arguable that if biological theory were ever to achieve the goal of constructing a rational system of forms, what *actually* happened in history would become of relatively minor interest. In physical science, once a theoretically grounded rational system of the elements had been established, the question of any historical relationships between the elements cannot have been regarded as one of great urgency since only relatively recently has the matter been seriously addressed.

The thrust of this argument, therefore, is to suggest that, in relation to the problem of form, the traditional dialectic, that between empirical classification and Darwinian explanatory theory, should be replaced by a new dialectic: a dialectic between a rational systematics and an explanatory theory of morphogenesis. I conclude by noting that, if a rational systematisation of forms in terms of transformations is possible at all, there is no reason why it should necessarily be restricted to relations between different individual organisms. In principle, it could be extended to relations between the parts of a single individual organism as Goethe anticipated. The grounds for this are outlined by Bateson (1894). He remarks that the Darwinists' preoccupation with relations of descent between individuals and their employment of the misleading metaphor of 'inheritance' as an explanatory concept has resulted in a sharp distinction being drawn between interorganismic and intraorganismic relations. On this view, there are two distinct types of 'homology'. This contrasts with the speculative thought of the early nineteenth century, in which individual organisms and individual members of a meristic series (for example, segments), together with the relations between these, were conceived as being alike in some sense. Harking back to this pre-Darwinist view, Bateson argues, on empirical grounds, that they might both be regarded as belonging within a single set of formal repetitions. Thus, he points out that, in many organisms, it is difficult to draw any clear distinction between metameric segmentation and asexual reproduction or between budding or strobilisation and colony formation. Asexual reproduction by division and the process by which ordinary meristic series are produced may be similar processes of repetition. If they are basically similar we may expect to find some analogy between the differentiation exhibited by the members of a meristic series and the variation between parent and offspring. And if we regard bilateral symmetry as a mode of repetition, then we might expect to find variation between the two sides analogous to that exhibited by two distinct individuals. Bateson claims that the empirical material fulfils all these expectations. I have drawn attention to the apparent similarity between intraorganismic developmental variations – homoeotic and heteromorphic 'transformations' – and interorganismic variations – genetic variations, phenocopies and terata. All appear to be specific.

While this purely empirical line of argument is by no means compelling, it is suggestive. In Bateson's view, it is possible that

the resemblance between the two sides of a bilaterally symmetrical body may be in some essentials the same as the resemblance between offspring of the same parent, or to use an inclusive expression, that the resemblance between the members of a Meristic Series may be essentially the same as the resemblance and relationship between the members of one family; that the members of a row of teeth in the jaw, of a row of peas in a pod, of a chain of Salps, or even a litter of pigs, all resulting alike from the processes of Division, may stand to each other in relationships which though different in degree may be the same in kind. (1894, p. 36)

Bateson thus envisages the possibility of a general science of biological form in which apparently disparate phenomena are unified by means of a theory of the generation of division, repetition and variation, that is, a theory of morphogenesis in the most comprehensive sense. As he remarks:

If reason shall appear hereafter for holding any such view as this, the result to the Study of Biology will be profound. (1894, p. 36)

Part II

Fields and Forms

My objective in this part of the book is to develop a theory of biological form that is based upon whole organisms as dynamically transforming systems that are technically described as fields. These are domains of spatial order, defined by internal relationships, that change in time according to well-defined principles or rules. This is the classical approach to problems of biological form, clearly described by Roux and Driesch despite their differences of emphasis, and developed by many biologists in this century such as Paul Weiss, Joseph Needham and C. H. Waddington. Weismann himself shared this perspective in his emphasis on development but, as Webster has shown, his separation of the organism into two distinct parts, an hereditary essence and a derived soma or body, had the effect of subjugating developmental processes to the effects of genetic causes.

This genetic reductionism needs to be examined closely in relation to developmental evidence to see if it can be sustained. In the first chapter of this part I take a look at the evidence that relates primarily to the formation of tetrapod limbs, since this is itself such a classical case of a fundamental biological form in which evolution and development are intimately intertwined. The issue is how to make sense of the developmental and evolutionary data in terms of generative principles – how limbs are made – and whether an understanding of gene activities is sufficient to achieve this goal. My conclusion is that gene action needs to be understood within the context of a theory of morphogenetic fields embodying organisational principles that themselves impose important constraints on the set of forms that can be generated. Genes are involved at every stage in the production of limbs in the developing organism; but the unity of structure that underlies tetrapod limbs as a category of biological form and defines them as homologous structures comes from the relational principles embodied in morphogenetic fields, not from the invariant action of genes. Without an understanding of these

129

organisational principles, which are emerging in models of limb morphogenesis, the different categories of biological form that are the foundations of taxonomy become unintelligible, accidental residues of historical events, as they are in Darwinism.

What emerges from this analysis is a theory of morphogenetic fields as complex dynamic systems that spontaneously undergo symmetry-breaking cascades: globally ordered initial fields pass through a series of bifurcations to detailed local structure that reflects initial order and results in an organism with coherently organised parts. The hierarchical nature of this generative process leads naturally to a hierarchical taxonomy of biological forms: ontogenesis provides the logical foundation for understanding phylogenesis.

Subsequent chapters elaborate on these themes by examining a variety of developmental systems selected to illustrate the universality of morphogenetic principles across the animal and plant kingdoms, and between multicellulars and unicellulars: segmentation in *Drosophila,* leaf arrangement (phyllotaxy) in higher plants and morphogenesis in unicellulars such as *Acetabularia* and ciliate protozoa. From these examples emerges the idea of morphogenesis as an intrinsically robust dynamic process that generates characteristic structures to which I give the name *generic forms.* These are the natural kinds that are revealed in evolution, the basic biological forms that are all transformations of one another under changes in the detailed dynamics of morphogenetic fields. At every level of any morphogenetic cascade there can be differences of dynamic detail, stabilised by genes in different lineages, and thus an immense (indefinite) diversity of final forms can arise. However, they all share generative principles that arise from the distinctive relational order of morphogenetic fields, and so there is a unity that underlies the diversity.

The implications of this view for our understanding of evolutionary processes are examined in the final chapter. Here explicit links are made between the approach explored in this book and the new field of study that goes under the name of the science of complexity, whose biological focus is an understanding of emergent phenomena in living systems from complex dynamic processes – that is, the emergence of biological forms. This rapidly developing interdisciplinary research area is now a major context for discussion and analysis of the issues raised in the present work.

6

Putting the Organism Together Again

The organism is, in a way, the paradigm structure. If we knew our own organism through and through it would, on account of its double role of complex physical object and originator of behaviour, give us the key to a general theory of structure.

Piaget, Structuralism (1971)

Molecular Reductionism and the Genetic Programme

Organisms have disappeared as real entities from biology. Their essences are identified with hereditary determinants, which Weismann located in a distinct part of the egg, the germ plasm. The information in these determinants, now identified with DNA sequences in the chromosomes, directs the development of the visible body of the organism from the rest of the egg substance, the somatoplasm. However, this body is transient and mortal whereas the hereditary essence is potentially immortal, being handed on from generation to generation by a lineage that is carried by the primordial germ cells, which give rise to the gametes for the next generation. As the self-reproducing hereditary determinants change by random variation, the soma changes and variants are produced on which natural selection can act. Weismann's dualism saved Darwinism from inconsistency by excluding the possibility of Lamarckian inheritance, and it provided a clear rationale for studying inheritance independently of development, since embryogenesis can be reduced to the activities of genes as the primary determinants of morphogenesis. They do this by producing the molecular materials out of which the body is made, different materials being produced in different parts of the developing organism so that structures specific to a particular type of organism arise under the direction of gene activity, resulting in an adult form distinctive to a species.

The metaphor used to describe this distinctive pattern of gene activities within a developing organism is that of a genetic programme encoded in the DNA. This is combined with the idea that the structural materials produced by the DNA (the molecules out of which the developing organism is made) have within them the specificity required to self-assemble into the different units of structure which constitute an organism of specific type. Self-assembly is familiar in all the molecular sciences, the paradigmatic example being atomic crystallisation. The clearest statement of this as the generative foundation of organismic structure is still that found in Monod's (1972) classic exposition, *Chance and Necessity*. Here we have the simple elegance of Lucretian atomism combined with characteristic Gallic eloquence. Although the same principles are used in current popular expositions, such as Dawkins's *The Selfish Gene* (1976) and *The Blind Watchmaker* (1986), the clarity of the basic principles become clouded in these treatments by ambiguities and qualifications that suggest the necessity for other generative principles but fail to identify them in clearly defined physical processes. The continuing power of genetic reductionism in biology is evident from the quotation from Delisi's (1988) paper that appeared in Chapter 4 and is repeated here:

[The] collection of chromosomes in the fertilized egg constitutes the complete set of instructions for development, determining the timing and details of the formation of the heart, the central nervous system, the immune system, and every other organ and tissue required for life.

Wolpert (1991) holds an equally reductionist position: "DNA provides the programme which controls the development of the embryo and brings about epigenesis."

Monod's text is very explicit about the causal logic involved in the view of organisms as genetically directed self-assembly systems. For him, they are "chemical machines" whose structure is generated by a process "strictly comparable to molecular crystallisation." He describes a hierarchy of these crystallisation processes:

1. Folding of the polypeptide sequences culminating in globular structures provided with stereospecific binding properties
2. Associative interactions between proteins (or between proteins and certain other constituents) so as to build cellular organelles
3. Interaction between cells, so as to constitute tissues and organs

Monod makes it quite clear that genetic determination requires that the structure generated at each level of the hierarchy be uniquely specified by the properties of the constituents at the lower level, so that any form is causally reducible to the primary protein structure of its constituents and hence to the historically given information in the nucleic acid sequence of the DNA. This is a basic principle of the genetic programme hypothesis, that the hereditary material of a species contains the information required to make an organism belonging to that species. Monod is, of course, aware that environmental conditions must be specified in order to realise a particular structure or form, but he restricts the definition of these constraints to variables such as temperature, aqueous phase and ionic composition. Let us see if these arguments are universally valid for biological processes, as Monod and others contend.

Although many examples now illustrate the same point, there is a classic case which illuminates the basic principles involved very clearly, and which antedates Monod's book by several years. This is the example of bacterial flagellum formation as described by Oosawa et al. (1966). These investigators showed that a homogeneous solution of the protomer, flagellin, obtained by the disaggregation of flagella from wild-type *Salmonella,* could be induced to reaggregate into either wavy (wild-type) or curly (mutant) flagella by adding to the solution wavy or curly seeds, respectively. These seeds, or nucleation centres, were fragments obtained by sonication of flagella of the two different types. Thus the same protein can, under identical environmental conditions, make different forms according to the way in which they are assembled under the direction of different crystal seeds, so that there is not a unique relationship between composition and form at this level of organisation.

Such a result is not in the least surprising, since the same type of polymorphism occurs in inorganic materials. One has only to think of diamond and graphite, or rhombic and monoclinic sulphur. The simple conclusion is that atomic or molecular composition does not, in general, determine form in crystallisation or self-assembly processes. Thus the molecular composition of organisms does not, in general, determine their form, even when this form can be reduced to pure self-assembly mechanisms. The state of any nucleation centres would also have to be specified.

Technically speaking, nucleation centres are the simplest examples of initial and boundary conditions on a dynamic process that generates a spatial pattern. The nucleating fragment of a flagellum defines the specific conditions on the boundary where polymerisation of the flagellum begins, specifying which of the possible spatial patterns is generated by the assembly process. When there is only one possible pattern, Monod's self-assembly hypothesis holds. However, in general there is more than one and the hypothesis fails.

At higher levels of biological organisation we find further examples contradicting Monod's proposition but conforming to what one expects of the more general process I have described, in which different initial and boundary conditions result in different spatial patterns generated by the same molecules. Again a classic example, well known when Monod wrote his book, is provided by Sonneborn's (1970) studies of the inheritance of altered form in the unicellular ciliate protozoan, *Paramecium.* Rotation of a ciliary row on the cell surface of an individual results in the appearance of a reversed row in all its progeny (observed up to the eight-hundredth generation). No change in the genetic information accompanied this heritable change of form, nor any change in the external environment, so that it is clear that more than one cell structure can be associated with a particular genotype. Sonneborn's (1970) interpretation of the observation is that the rotated ciliary row acts as a local nucleation centre for the assembly of gene products during preparation for cell division. Hence the altered form is transmitted from generation to generation by a mechanism like that involved in flagellin assembly into flagella, but now at a higher level of organisation, the ciliary row, which is a basically two-dimensional structure. Again it is clear that the specification of form requires more than the specification of the genotype, thus falsifying genetic determinism. We conclude that understanding how an organism of a particular form is produced requires a knowledge of more than its molecular composition and the environmental conditions.

It is sometimes objected that examples from the ciliate protozoa are not of general biological validity because the type of nongenic inheritance involved is dependent upon continuity of the soma during asexual reproduction, which is not the case in sexually reproducing species. There are two points to be made about this. First, any comprehensive theory of biological inheritance must include processes common to many asexually reproducing species. That nongenic inheritance is not confined to unicellulars was demonstrated by Sonneborn (1970) with the example of the turbellarian worm, *Stenostomum,* which showed the same type of inheritance of acquired form as *Paramecium.* He produced face to face and back to back doublet organisms, which reproduced true to their new forms. Secondly, what is at issue is not just mechanisms of inheritance, but what needs to be known in order to understand how organisms are made. What happens in asexual reproduction is fully relevant to all developing organisms, for it is the development of the soma that concerns us here.

When Monod's third level of self-assembly is considered, the interactions between cells of specific type carrying specific surface markers, the refutation of atomism depends upon less direct inferences from experimental evidence than those relevant to molecular self-assembly. Suggestive studies of the type carried out by Zwilling (1964) provide an instructive beginning. He took limb-bud mesenchyme from Stage 19–20 chick embryos, disaggregated the tissue into cells, pelleted them by centrifugation and replaced them under limb ectoderm. No limbs of typical form were produced, although limblike structures of diverse morphology were frequently formed.

Similar types of experiment on chick limb buds were also conducted by MacCabe, Saunders and Picket (1973), with similar results. After spatially scrambling limb mesenchyme cells and placing them under an ectodermal jacket, a diversity of limblike structures were formed with mesenchymal condensations resulting in typical cartilage-generating sites arranged in branching, segmental arrays characteristic of basic limb skeletal patterns. However, many of the skeletal elements produced could not be identified as normal limb elements, having intermediate and transformed characters. These examples show that limb mesenchyme cells can form structures different from a normal limb, suggesting that cellular interactions alone are insufficient to specify an unique morphology. The inference is once again that the generation of a particular structure in an organism depends not only on the properties of the elements making up the structure (cells in this case) but upon additional influences affecting the spatial order that emerges from cell–cell interactions.

In the case of the chick limb, one of these influences has been well characterised. It was discovered by Balcuns, Gasseling and Saunders (1970) that tissue at the posterior margin of the limb bud of the chick exerts a spatial organising influence on the developing limb, generating the asymmetry of the antero–posterior axis. This is known as the zone of polarising activity (ZPA). Similar regions, playing a similar role, have been identified in other species of bird (Fallon and Crosby, 1977), in a reptile (alligator, Honig, 1984) and in amphibians (Cameron and Fallon, 1977; Slack, 1976), while polarising zones from both mouse and snapping turtle limb buds have a polarising effect on chick limb buds. The ZPA seems to be a tetrapod universal. Its influence extends from the posterior boundary of the limb bud, producing a characteristic antero–posterior asymmetry in the developing limb. When the mesenchyme, including the ZPA, is spatially scrambled and replaced under the ectoderm where it can express its limb-forming potential, a different structure from the normal is produced because the long-range spatial patterning influences that normally act have been disturbed.

These arguments are all very familiar to developmental biologists, and it is widely accepted that the limb field, the domain of tissue within the flank of the embryo competent to produce a limb, includes influences such as those coming from the ZPA. Research on this system has now progressed to the molecular level, with an active agent being identified as retinoic acid (Maden,

1982a, 1983, Tickle et al., 1982, Maden and Keeble, 1987), while genes of the hedgehog family, involved in determining polarity of developing structures in both invertebrates and vertebrates, are also implicated (Heemskerk and DiNardo, 1994; Smith, 1994). A vigorous investigation is proceeding in a number of laboratories on the mechanism of their action. In relation to atomistic reductionism, the relevance of this example is the demonstration that, as in previous cases, the particular spatial structure that is generated depends upon both the properties of the assembling or aggregating elements (the mesenchyme cells in this case) and spatial patterning influences coming from the boundaries of the field and producing long-range order. So a strict atomism of the self-assembly type proposed by Monod is untenable at all the levels described by him. But there are probably few biologists who would now subscribe to such a pure form of atomistic reductionism or "molecular ontogenesis," as Monod called his analysis of morphogenesis. It is generally accepted that fields with initial and boundary conditions involving long-range patterning effects are the generators of specific morphologies. However, it is possible to accept the occurrence of such influences in developing organisms and still adhere to the genetic programme metaphor, as we shall see in the next section.

Positional Information and Interpretation

From the foregoing analysis we see that the attempt to articulate Weismann's concept of 'determinants' in the germ plasm causally responsible for the generation of organisms in terms of genetic instructions and molecular self-assembly has not been successful. However, many models have been advanced which propose various additions to this scheme in an attempt to provide sufficient information to give an adequate morphogenetic theory. These all involve postulates about spatial information in the organism in the form of segregated cytoplasms, gradients of 'morphogens' or unspecified determinants which, together with the genetic program are intended to provide a complete specification of the morphogenetic process. One such theory which has been particularly influential is that developed by Wolpert (1969, 1971). In his theory of positional information, Wolpert assumed the absolute minimum required for spatial order in a developing domain, namely a coordinate system which assigned a unique positional label to every point within it. The boundaries of the domain were defined by extremal values (maxima or minima) of the coordinate system, generated by a mechanism which was not specified but was assumed to be based upon chemical reactions producing 'morphogens' whose diffusion through the domain produced spatial continuity of state in the form of gradients. This minimal degree of spatial order represents a significant advance on a simple local molecular self-assembly scheme. A cell located at any point within such a monotonic gradient was then characterised by a unique positional value specified by one, two or three morphogens according to the dimensionality of the domain. This provided positional information which the cell's genome then 'interpreted' in a manner specific to the species and to the cell's developmental history, the interpretational process being described as the production of a particular pattern of gene activities that generate the final differentiated form of the cells in a particular position of the embryo, hence causing morphogenesis.

Wolpert realized that by locating all the specificity of positional response in the genome, the coordinate system or positional information could be universal: That is, every distinct domain of a developing organism could be specified by the same type of coordinate system, since the particularities of any pattern arise from the interpretation process coded in the genes. Furthermore, this universality could be extended over all organisms. Universality thus became a basic postulate of the theory and was used by Wolpert to explain a variety of phenomena observed in developing organisms.

A further elaboration of the theory of Wolpert and Lewis (1975) contains the following statement: "A theory of development would effectively enable one to compute the adult organism from the genetic information in the egg." The theory is thus thoroughly Weismannist in conception, being based upon a sharp distinction between an 'active' decision-making genome ('germ plasm') and a 'passive' nonspecific coordinate system ('somatoplasm'). Wolpert (1969) is explicit that there is no relationship between the pattern of positional information and the specific morphology which is generated by the genomes within it. His solution of the problem of morphogenesis in the individual organism, conceived as the spatially patterned determination of cell states, is thus formally identical to Weismann's solution of the problem of temporal and regional polymorphism and employs the same conceptual scheme. Weismann argued that a temporally or spatially organised variable in the external environment selected which particular set of determinants in the germ plasm would be expressed, and hence which of the possible alternative forms would be produced. Wolpert argues that a spatially organised variable in the internal environment ('positional information') determines which particular set of genes in each member of a (necessarily) spatially distributed set of genomes will be expressed and hence, ultimately, which component part of the total form will be realised in that part of embryonic space.

Wolpert's scheme is Weismannist in reliance on the historically given 'central directing agency' as the determinant of form, and morphological diversity is inevitably, therefore, seen as irreducible. There can be no general laws of form or generative principles in such a theory since the only universal, positional information, imposes no constraint upon the form which is generated other than that it be spatially extended. The question arises whether this form of Weismannism is consistent with the empirical evidence. I now argue that it is not.

Positional Information and Prepatterns

Let us return to the chick limb, the developmental system that has been so extensively used by Wolpert and others to assess the validity of his theory. The initial application of the model to the specification of pattern across the A–P axis involved postulating the existence of a monotonic gradient in a 'morphogen' that had a genetically specified maximum value at one boundary of the limb field and a minimum at the other. The different elements of the limb are then generated by discrete concentration ranges of the morphogen, well-defined threshold points being specified by the genetic programme to which each cell has access in its DNA. Cells finding themselves within a particular range of morphogen 'interpret' the positional information in such a way as to produce the components of the limb element in that position. With such a mechanism, it is unnecessary to have any similarity of spatial pattern (isomorphism) between the morphogen and the overt structure generated by it. The different proximo–distal regions of the limb where there are one, two or three elements simply have the appropriate number of threshold points (0, 1 or 2, respectively) appropriately positioned in the same gradient across the A–P axis, these thresholds being specified by the genetic programme.

This lack of isomorphism between the pattern of morphogens and the structure generated was criticised (Goodwin, 1976; Goodwin and Trainor, 1983) for its genetic and dynamic implausibility and its inconsistency with the evidence. The work of Zwilling (1964) and of MacCabe et al. (1973) on the capacity of scrambled mesenchyme to generate limblike structures argues against the necessity for any monotonic gradient in primary pattern formation in the limb, since such a gradient would be abolished in homogenised mesenchyme aggregates.

Further evidence of the fundamental capacity of tetrapod limb mesenchyme to spontaneously generate discrete domains of aggregated cells resembling the prechondrogenic condensations of the normal limb skeleton are provided by Pautou's (1973) experiments in which disaggregated duck and chick leg mesoderm cells were mixed and repacked under an ectodermal jacket. These

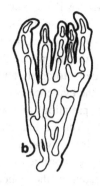

Figure 6.1. Limbs formed from disaggregated mesoderm with 4 (a) and 5 (b) digits that lack anteroposterior asymmetry.

formed skeletons starting proximally with a single element which branched and segmented to generate a pattern of four or five atypical digits, lacking any antero–posterior asymmetry (Figure 6.1). This evidence all argues against determination of multiple periodic elements, such as digits, by monotonic morphogen gradients. It is much more plausible to assume that the primary pattern generator produces a 'prepattern' which is isomorphic with that of the overt structure, as in fact was originally proposed by Turing (1952), who introduced the term 'morphogen'. Three digits, for example, imply a prepattern with three spatial periods. An early application of this idea to the case of the chick limb was presented by Wilby and Ede (1976) and by Newman and Frisch (1979).

Interpretation and Naming: Darwinian Homology

Wolpert now accepts these arguments and has adopted the position that morphogen prepatterns in the limb are most likely to be isomorphic with the basic pattern of the structures generated (Wolpert and Stein, 1984). What then remains of positional information is the process of 'naming' the elements, of distinguishing them from one another. In the case of digit specification, this is assumed to be the role of the morphogen arising from the ZPA, together with particular genes involved in this process.

The job previously achieved by a single morphogen in organizing the antero–posterior characteristics of the limb is now carried out by two. The first one, distributed in a periodic spatial pattern isomorphic with the distribution of condensation domains, on its own gives a basic limblike form of the type shown in Figure 6.1b. The second, distributed in a monotonic gradient from the ZPA, is involved in imposing particular characters or identities on the chondrogenic sites as they develop, resulting in the digit designations 2, 3 and 4. (Although the convention is to use Roman numerals for the 'names' of digits, I shall use Arabic numerals throughout.) One might wonder at this point why chick forelimb digits are so named, rather than 1, 2 and 3 or A, B and C. Indeed there is continuing dispute about this, the embryological evidence suggesting 2, 3, 4 while palaeontologists tend to accept the designation 1, 2, 3. A recent discussion of the evidence is given by Shubin (1994). The dispute is based on historical interpretations of homology, reflecting precisely the concepts of descent and inheritance introduced by Darwin as explanatory principles of biological form, whose limitations have been examined by Webster in Chapter 3. They continue to exist in modern biology, descending from Darwin via Weismann to the genetic programme as the vehicle of historical determinants of morphology. The argument in relation to limbs is as follows.

The 'original' tetrapod limb, that which belonged to the common ancestor of the four-legged lineage, is assumed to have been one with five digits, since this is the condition of primitive, extinct amphibians such as *Eryops* and continues to be the predominant amphibian limb form. These digits were accordingly 'named' 1, 2, 3, 4 and 5. The Darwinian belief that there is some type of inherited, material continuity of the individual elements that constitute a persisting form, such as the tetrapod limb, results in the view that the elements in the limbs of different species can be named in relation to the original digits. Thus, despite the fact that there are very considerable differences of form in chick phalanges (Figure 6.2) compared with those of amphibians (Figure 6.3), the general similarities of structure such as those defined by the digit formula (number of phalanges per digit), relative length and position are used to make these identifications. The conclusion is that the chick has effectively preserved digits 2, 3 and 4 of the common ancestral limb, albeit greatly altered by an accumulation of small changes during evolution. The most extreme example of this historical modification of a postulated ancestral pattern is to be found in the limbs of ungulates, such as the horse. Here a single digit, identified as 3, has become very large, while the others (2 and 4) are reduced to tiny elements.

These are classic examples of the Darwinian concept of homology; similarities of morphological structure in different species within a common lineage are to be explained by descent from a common ancestor. The eminent Darwinian, Sir Gavin de Beer, examined the implications of such a conception and found them incoherent, as described in his fascinating little treatise *Homology: An Unsolved Problem* (1971). He points out that the similarities of form between the fore and hind limbs of any species "is not real homology, as fore-limb and hind-limb cannot be traced back to any ancestor with

Figure 6.2. The structure of a normal chick fore limb.

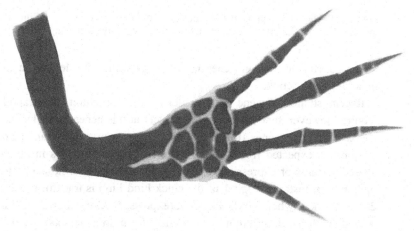

Figure 6.3. The structure of a normal amphibian fore limb (*Ambystoma mexicanum*).

a single pair of limbs.'' In fact, the evidence is that fore limbs and hind limbs evolved independently from pectoral and pelvic fins in primitive gnathostomes. This leads us to the conclusion that human arms, bats' wings and the fore limbs of a horse are homologous, but human arms and legs are not. This is not a useful result, so it is necessary to seek a more satisfactory definition. de Beer then enquires whether there is a genetic basis for homology, homologous structures involving the action of the same genes. However, the mutation of a single gene can cause morphological changes in characters that are not homologous, pleiotropy being the hall-mark of gene action. Furthermore, identity of morphology of particular parts of different individuals does not imply similarity of genotypes, which de Beer illustrates by the case of the *eyeless* mutation in *Drosophila*, already mentioned by Webster in Chapter 4. This mutant gene results in a failure of eyes to form, but normal eye morphology returns after several generations as a result of changes in other genes, despite the continuous absence of the *eyeless*[+] gene. So we are forced

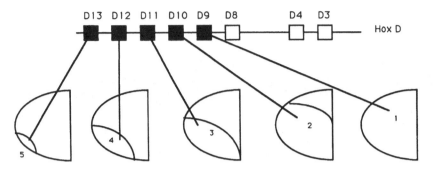

Figure 6.4. The overlapping spatial patterns of the Hox genes active in limb buds and their relations to the different digits defining pentadactyly, according to Tabin (1992).

to the conclusion that, in general, homology cannot be defined in terms of invariant gene action.

Recent studies of gene activity during the formation of tetrapod limbs require, however, that we look more closely at this general conclusion. Homeobox-containing genes (the HoxD cluster) are active in tetrapod limb formation. As expected from the mutant effects of such genes in *Drosophila*, altered patterns of expression in these genes can result in homeotic transformations: for instance, digit 1 of the chick hind limb is transformed to a digit 2–like element when HoxD 11 is overexpressed (Morgan et al., 1992). This accords with the description of the digits of a limb as an example of ''serial homology,'' which Bateson used for repeated but not identical parts of an organism.

HoxD 11 is one of five genes involved in the formation of the tetrapod limbs that have been examined so far. Tabin (1992) suggested that they are expressed in a spatial pattern that could provide unique combinations for each of the five digits, as shown in Figure 6.4. He presented the argument that pentadactyly is to be identified with the constraint imposed by this combinatorial genetic code: there cannot be more than five distinct digits. If limbs have more than this number, as they do in conditions of polydactyly and in fossil tetrapods (Coates and Clack, 1990), then digit characters must be duplicated, as Tabin suggested for the case of the seven digits of *Acanthostega*. This describes a genetic constraint, an historical invariant that is identified with a particular spatial pattern of gene transcriptions. The question arises: does this lead us to a general definition of homology in terms of invariant gene action? Hinchliffe (1994) has expressed the following reservations. ''Implicit in much of the recent emphasis on homeobox expression domains encoding positional address is the belief in a simple relation between genes and morphology, exemplified by the five-domains–five-digit hypothesis. But, in my view, the development of structure cannot be fragmented in this way. Parts of the skeleton are formed by co-related cellular processes and inductions organised both in time and in space.'' Experiments have now revealed discrepancies that make Tabin's proposal unlikely, and he has himself altered his position (Morgan and Tabin, 1994). However, it is worth pursuing this example further because this general type of argument will recur in attempts to explain homology in terms of invariant relations between patterns of genes and morphology. I shall now point to what I see as the basic inconsistencies of this interpretation and suggest that the type of constraint involved here finds its place naturally within another conception of homology which is closely related to Bateson's way of describing it.

Homology as Equivalence

As Webster has discussed (Chapter 3), Bateson criticised an analysis based on a common ancestral form on the grounds that there is an inconsistency of interpretation between cases of repeated and more or less identical structures (e.g., intestinal segments of the earthworm, teeth of a roach, petals of a flower or spinal vertebrae) and those in which there are individual differences, as in the case of digits. It is futile, he argued, to attempt to homologize individual vertebrae between species. Rather, what one sees is a variation in the number of elements between species and secondary modifications that result in a graded series of changes between elements of the series. Then the individual elements have no persistent identity between species, and the use of names to imply such an identity is misleading and basically erroneous. Bateson's position was that there is indeed an intelligible unity of process underlying meristic series in different species, but that this is not to be found at the level of individual elements and their identities. Rather, he implies that this unity is in the generative process that establishes the members of the series. Differences between them result from asymmetries such as those produced by the ZPA. So what is now needed is a dynamic definition of homology as generative process, not as final pattern.

I shall define homology as an equivalence relation over the members of a set, defined by a transformation that takes any member into any other member within the set (see also Goodwin, 1982). This concept of equivalence under transformation is commonly used in mathematics. For example, we can consider the set of shapes produced by transformations that preserve connectivity and so do not introduce or eliminate holes in some initially given shape. A sphere can be distorted into any shape that preserves the connectedness of all points in the sphere; or a doughnut (a torus) can be transformed into any shape that conserves the hole. This class of transformations defines topological equivalence. The invariant is simply connectedness. The mappings can obey the constraint that the angles between intersecting lines in one space are preserved in the image space. These define equivalence under conformal mappings. Or distances can be preserved, as in the motions of rigid bodies (translations, rotations and inversions), giving equivalence under metric transformations. Each equivalence set has an invariant property that is preserved by the transformation. A hierarchy of such equivalence classes can be defined, each successive invariant including but imposing more constraints than the previous one.*

Developmental and Taxonomic Hierarchies

Developmental processes are hierarchical. So are biological classification schemes. Their relationships will be examined as different examples are discussed, taking us on a route towards a theory of biological forms and their logical relationships (a rational taxonomy). Considering first tetrapod limbs, let us start with the highest level of the hierarchy, the organisation of the whole limb, and proceed downwards to smaller units, which will bring us back to digits. Classical embryology established the limb field as the developmental unit of spatio-temporal organization at this level (Harrison, 1921), although of course the limb field is itself initiated within a larger field, the whole embryo. Its relative autonomy comes from the observations that, at a particular stage of development, a limb field can be transplanted to an ectopic site in the embryo and produce a supernumerary limb.

The next level of the limb-generating process is the production of the basic pattern of the limb elements. The first definition of their invariant relational properties was given by Geoffroy St Hilaire in terms of his Principle of Connections (see Chapter 2 of this book, The Concept of Type). We now

---

* Lewis and Wolpert (1976) used the term 'equivalent' in a quite different sense, meaning identical, in their definition of nonequivalence as applied to the difference between, for example, fore and hind limbs. They find it necessary to emphasise that such structures are not morphologically identical despite the fact that they are made from the same cell types – chondrocytes, myoblasts, osteoblasts, fibroblasts, etc., which are not themselves identifiable as belonging to fore or hind limbs. Such an emphasis is required only if one takes a reductionist view of Monod type that identity at one level implies identity at another. However, morphogenetic fields can clearly differ in ways that are not reducible to cell type, such as initial and boundary conditions in variables associated with mechanical properties (e.g., strain along different dimensions) and the state of the extracellular matrix as well as the concentrations of different morphogens. So this notion of nonequivalence is not necessary in a field description of morphogenesis.

need a dynamic description of this process and an identification of whatever constraints may be operating. It used to be thought that all limb buds initially generated the archetypal or ancestral pentadactyl limb pattern, subsequently undergoing transformation to the different tetrapod forms by secondary modifications such as fusion. This is now known to be incorrect. In the chick, for example, the number and relations of the chondrogenic condensation sites in the limb-bud mesenchyme are the same as the elements of the adult limb (Hinchliffe and Hecht, 1984; Hinchliffe, 1990). So we need to examine this condensation process in order to discover both the constraints that are responsible for the regularity of tetrapod limbs and the variations within these constraints that result in the range of possible forms that constitute the transformation set.

At this point it becomes necessary to use a model of the condensation process that suggests what the origin and nature of the constraints might be. This is because the enterprise that we are engaged in is theoretical in the sense that the set of possible tetrapod limb forms is to be understood in terms of a generator whose rules of operation can be described and used to actually generate the patterns representing the members of the transformation set. This is the modelling exercise and it proceeds tentatively, in dialogue with experimental study.

The model that I shall now make use of to illustrate the argument is that proposed by Shubin and Alberch (1986) and by Oster et al. (Oster, Murray and Maini, 1985, and Oster et al. 1988) to describe the mechanics of the mesenchyme aggregation process that gives rise to the elements of the developing limb. This is not a complete description of limb formation and it will undoubtedly turn out to be wrong in certain respects, like all models, but it provides a very good focus for the issues involved in explaining morphogenesis. The model is based upon experimental studies of the behaviour of mesenchyme cells during their condensation to produce the cartilaginous elements that later become the bones of the limb, and on a mathematical model that demonstrates how these processes could occur by plausible mechanisms of cell adhesion and the behaviour of aggregates as they grow (Oster, Murray and Harris, 1983). In the model, condensation is initiated by a local change in the osmotic properties of the extracellular matrix which results in its collapse, bringing cells closer together. Adhesive forces between cells then result in a focus of cell aggregation. This grows by recruitment of cells which become polarized along the tension lines in the ECM and move towards the aggregate. Their movement generates traction forces that reinforce polarity and directed movement. During growth of the condensate, whose direction is partly dependent upon the geometry of the limb bud and by growth along the proximo–distal axis of the limb, the condensation process can become unstable and bifurcate in one of only two ways. It can branch to produce a Y-shaped structure; or it can segment, producing another discrete condensation that is on the same axis as the first (see Figure 6.5). The arrangement of elements in a limb is then a result of these processes: initial or focal condensation, branching bifurcation and segmentation. These are suggested as the primary components of the generative process at the level of limb element production, according to the model. They explain well not only normal limb patterns but also structures such as those shown in Figure 6.1. We can thus recognise different levels in the hierarchy of limb-forming processes. These include (1) individuation of the limb field as an autonomous domain; (2) establishment of axial order within the limb field; (3) formation of the limb elements by condensation, branching and segmentation; and (4) modification of the elements along the A–P axis by influences such as those deriving from the ZPA, *sonic hedgehog* and HoxD genes. These overlap in time, but can be distinguished within the unitary process that is the unfolding of spatial order in the limb field.

Naming versus Transformation Sets

Consider now the influences such as those of the ZPA on limb-bud aggregates. There are essentially two different ways of interpreting such action, with distinct experimental consequences. The first of these preserves aspects of Darwin's historical perspective and the notion of a continuity of recognisable elements throughout the tetrapod lineage by describing the action of the ZPA as a 'naming' process that confers separate identities on the elements in accordance with a genetic programme that is distinct to the species. This is the position of Wolpert and Stein (1984), who insist that "the naming is a discrete process and no digits of intermediate character are ever formed as would be expected if, instead of naming digits, the polarizing region were merely altering growth parameters in some continuous way." That is to say, the transformation set available to the elements of a limb in any species is restricted to the normal set of characters specified by the genes of that species. The possible forms of organisms are thus defined by genes. But since there is an indefinite set of possible genes, there is an indefinite set of possible forms and there are no properties of the generative process that impose logical constraints on this set. Biological forms are therefore unintelligible and all taxonomies must be arbitrary classifications of historical accidents. This is the Darwinian position, whose implications have been examined by Webster

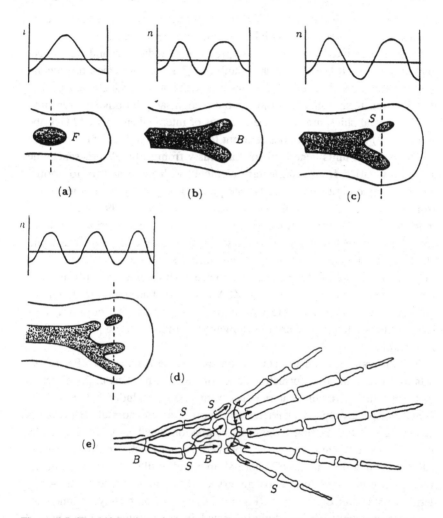

Figure 6.5. The bifurcations that generate limb elements in the model of Oster et al. (1988): (a) focal condensation; (b) branching bifurcation; (c) segmentation, with an interpretation applied to an amphibian limb (e). State profiles showing nodes or waves across the antero–posterior axis defined by the dotted lines are shown in the graphs. (Reprinted from Murray "Mathematical Biology," *Biomathematics.* Vol. 19, © Springer-Verlag, 1989 Fig. 17.16, p. 565)

in Part I. The alternative to naming is for the ZPA to effect a variation of parameters influencing the condensation process such that a set of digits with characters different from the normal could be generated by experimental disturbance: that is, a transformation set would exist which includes the normal forms together with structures of mixed or intermediate character, but constrained by the generative principles of morphogenetic fields.

These alternatives have clear-cut experimental predictions and are subject to empirical test. It is evident that scrambled chick mesenchyme can generate digitlike structures that cannot be identified with normal limb elements. Does the ZPA transform such indeterminate but recognisably limblike patterns into species-typical chick limbs, or does a series of intermediates arise? MacCabe et al. (1973) made observations relevant to this question. First, they reported that the skeletal units generated spontaneously from scrambled mesenchyme aggregates varied from rudimentary elements with no similarity to normal structures up to digits that could be interpreted as digits 3 and 4 of the normal wing complement, though the latter was very rare (2 of 99 cases). When fragments of ZPA were implanted into the anterior or posterior margin of scrambled mesenchyme aggregates, a series of patterned elements was again obtained, extending from ones bearing little resemblance to normal limbs (now in the minority) to those that allowed unambiguous identification of limb elements and digits, although ZPA rescue of the randomized cell aggregates never produced more than two digits. The digits were always arranged with an antero–posterior asymmetry determined by the position of the ZPA implant.

It is difficult to avoid the conclusion from these results that ZPA rescue acts by shifting the transformation series of limblike patterns in the direction of normal limb structure, intermediate forms being included between those that lack species-typical character and those that are normal. However, a crucial test of the naming hypothesis centres on whether or not single elements of mixed character can be obtained. In the chick, grafting a second ZPA to the anterior margin of a normal limb bud results in mirror-symmetric limbs, since now the normal asymmetry of structure across the antero–posterior axis is mirror reflected (Figure 6.6). In a completely symmetrised structure, not only are the digits reflected in the midline, but the normal asymmetry of the lower arm (radius and ulna) is replaced by two mirror-imaged ulnae. However, different experiments result in different degrees of transformation, and Summerbell (1981) has observed examples in which a single element has a proximal epiphysis that is quite clearly that of a radius, while the distal epiphysis of the same element is definitely that of an ulna (Figure 6.7). This is clear evidence for transformations and against interpretation of positional information as a naming process of the type defined by Wolpert and Stein (1984).

Wolpert now (personal communication) takes the view that such an ele-

Figure 6.6. A mirror-imaged chick fore limb resulting from an anterior ZPA graft. (Courtesy of D. Summerbell.)

Figure 6.7. An incompletely mirror-imaged chick fore limb in which one element is a composite of radius and ulna (upper element of the mid-limb). (Courtesy of D. Summerbell.)

ment of mixed character is consistent with the naming hypothesis. What he predicts is that there will be a sharp discontinuity of molecular labels at some point along the element, where a threshold is located. This implies that 'naming' is a molecular process in which specific genes are activated by specific concentration ranges of a graded morphogen, but a discontinuity of molecular labels need not result in any morphological discontinuity in the generated structure. The hypothesis that 'interpretation' acts upon single elements of structure and 'names' them according to a limited set of discrete possibilities determined by the genotype is thus altered. The naming process can now operate on an element of structure generated by an earlier process (e.g., the prepattern generator that produced the single element that later acquired mixed radius–ulna identity) and give it any identity, so long as this is a discrete mixture or mosaic of adult characters, though there need not be any structural discontinuities in this mosaic.

The distinction between positional information and interpretation as the process that 'names' or gives identity to the elements of adult structure thus acquires an arbitrary and ad hoc flavour. The evidence from limb studies suggests that the duality of Wolpert's theory does not provide a useful conceptual framework for understanding how the morphological elements of adult structure arise. Gene products contribute to the physical processes of morphogenesis throughout the whole of epigenesis, from the individuation of the limb field to the condensation of mesenchyme and the emergence of skeletal elements of characteristic form. There is no specific stage in this process or set of gene activities that can be described as 'naming' the structures generated. Elements of mixed character can arise from variation of the influences (genetic or environmental) acting through the morphogenetic field at any level of the process, resulting in a set of possible forms which are equivalent under transformation of the generative process. A major objective of Wolpert's dualism is to preserve the primacy of genetic factors as the essential determinants of biological form so that epigenesis reduces to a genetic programme. As observed earlier, this dualism is essentially equivalent to Weissman's, from which it originates. But, as pointed out by Webster, a particular set of genes is neither necessary nor sufficient to generate a particular structure and therefore no invariant relationship can be defined between genes and form. A conceptual reformulation of the relationship between gene activities and morphogenesis is necessary in order to clarify the nature of the causation that underlies the emergence of biological form, which is a primary objective of the present work. To illustrate the logical difficulties of the view that patterns of gene activity can be used to define and classify morphological relationships, let us return to Tabin's proposal concerning genetic specification of digits via the HoxD genes.

Genes, Fields and Homology

Specific spatial transcription patterns of HoxD and other genes across the limb bud influence the detailed structure of the digits characteristic of different species. Furthermore, within a species digit characters can undergo transformation one into the other by alterations of HoxD gene expression (Morgan et al., 1992), so digits can be homologised. But, according to the definition of homology as equivalence, it is not clear that the digits that have been named 3, for example, are homologous across different species. What are the generative invariants that identify what is called digit 3 as distinct from digit 2, say? In the cat hind limb there are four fully formed toes, usually designated 2, 3, 4 and 5, each with three phalanges. Digit 1 is rudimentary, consisting of a small cylindrical bone. In the chick hind limb, digit 1 has two phalanges, digit 2 has three and digits 3 and 4 each have four phalanges. Similar variations can be found across other species. What is it that identifies digits 3 as a homologous set that excludes the other digits? It is not at all clear that there is an identifiable constraint that allows us to designate these digits as a distinct homology class (see also Goodwin and Trainor, 1983; Goodwin, 1994b). The action of the HoxD genes does not provide this evidence. In fact, the Hox combination proposed by Tabin (1992) that corresponds to digit 1 in the chick hind limb (D9 +

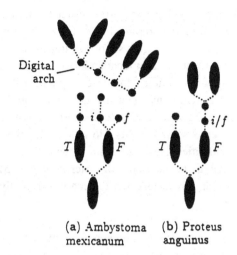

Figure 6.8. A comparison of the proposed generative sequences of the elements in the limbs of two different amphibians (a) *Ambystoma mexicanum* and (b) *Proteus anguinas*. (Reprinted from Murray "Mathematical Biology," *Biomathematics*. Vol. 19, © Springer-Verlag, 1989 Fig. 18.10, p. 607)

(a) Ambystoma mexicanum   (b) Proteus anguinus

D10) is not the same as that which corresponds to digit 1 in a five-digit amphibian hind limb (D9), so Hox gene combinations do not define digit identity or homology in such a scheme either.

Experimental studies of digit loss of the type presented by Alberch and Gale (1985), using mitotic inhibitors, also fail to resolve digit identities. Their interesting observations suggested that evolutionary digit reduction occurred from the posterior (postaxial) region of the limb in certain taxa and preaxially in others. However, digit reduction is likely to be accompanied by change in the generative process. An example of this is provided by Figure 6.8, which is the interpretation presented by Shubin and Alberch (1986) of the generative processes giving rise to the hind limb patterns of the salamanders *Ambystoma mexicanum* and *Proteus anguinus*. The latter shows extreme digit reduction, and the branching bifurcation that is proposed to give rise to the intermedium (i) and the fibulare (f) from the fibula (F) in *Ambystoma* is replaced by a segmentation into a single element (i/f) in *Proteus*. What is the 'name' of this element in relation to the elements of an ancestral limb? And what is the correct designation of the digits? Are they 3 and 4 or 4 and 5? Clearly these are not answerable questions. As stated by Murray (1989): "It is dangerous to relate geometrically similar elements without knowledge of the underlying developmental programme, for the processes that created the elements may not correspond. . . . it is not sensible to ask 'which' digit was lost, since the basic sequence has been altered." Wagner (1994) adds to this view with the observation that phylogenetic homology has been superceded by cladistic concepts, which allow one to pursue genealogical (historical) issues by using history-free criteria. The task now is to concentrate on what Wagner calls biological homology (i.e., generative regularities).

A consistent analysis of the transformational relationships between forms depends upon an understanding of the causal mechanisms that operate in the limb field, defining the necessary and sufficient conditions for generating the forms. These mechanisms are contingent (historical) in the sense that particular substances and processes are required to realise the generative activities of tetrapod limb fields. However, once these fields have arisen, they express regularities that come from certain invariant properties of their dynamic behaviour. In the model of Oster et al. (1985), these are the three bifurcations that are characteristic of (generic to) the dynamics of the condensation process.

Tetrapod limbs evolved from fish fins by a process involving changes in the generative mechanisms that gave rise to a remarkable diversity of groups undergoing similar types of transformation, with pentadactyly probably emerging independently in different lineages (Coates, 1994). There are essentially two aspects to this evolution: the loss of the distal exoskeletal elements (the fin ray) and the increasing complexity of the endoskeleton. Thorogood (1991) has proposed that the former resulted from a heterochronic shift, the generative period of the endoskeleton (which forms first) being extended and that of the exoskeleton being delayed and abbreviated until it disappeared entirely. The evolution of the endoskeleton involves an elaboration of the bifurcation mechanism from the condensation and segmentation processes that generate the bilaterally symmetrical patterns of fish fins, to the additional branching bifurcations and the development of broken symmetry across the antero–posterior axis by the influences from the ZPA and the HoxD genes that characterise tetrapod limbs. Homological relationships are then defined by the similarities and the differences of generative process between fins and limbs, independently of lineage, allowing one to construct a rational taxonomy in terms of causal processes and their transformations. Strictly speaking, a polyphyletic origin of pentadactyly means that such limbs arising in different lineages are not homologous in the Darwinian sense, even if they involve the same generative mechanisms. This is not a morphologically useful position to adopt.

The arguments of this chapter lead us to the conclusion that the dualism of Wolpert's theory, which is essentially the same as that of Weismann, is difficult to reconcile with the evidence and does not provide a conceptual framework within which biological forms and their evolutionary transformations can be understood. There are not separable processes of spatial patterning and interpretation, the genotype naming the elements generated by a prepattern and so giving them specific character. Nor are genes the causes of epigenesis, as described by a genetic programme. The dynamics of development involves a different kind of process in which gene action occurs within the context of organised processes with distinctive causal powers, morphogenetic fields. Elaborating on the nature of these fields is what the rest of this book is about.

7

Segments, Symmetries and Epigenetic Maps

Such, then, is Symmetry, a character whose presence among organisms approaches to universality.

William Bateson, "Materials for the Study of Variation" (1894)

Mutations and Mirror Symmetries

The segmented, bilaterally symmetric structure of *Drosophila* is representative of one of the basic animal body plans. Segments provide a classic case of a meristic series, each with identifiable characteristics but transformable one into the other by either genetic mutation, as in the homoeotics, or by environmental perturbation, resulting in so-called phenocopies. There is now an overwhelming wealth of data on morphological mutants and the spatial patterns of gene products involved in segmentation in *Drosophila* embryos, so that this fundamental process is now visible in unprecedented molecular detail, as described in Lawrence's (1992) comprehensive monograph. The challenge is to make sense of what has been revealed. In this chapter, the objective is to make use of particular aspects of this process that bear the marks of generic properties, pointing to quite deep organisational principles of epigenesis. These can then be used to carry the arguments of previous chapters in the direction of a coherent picture that not only addresses basic issues of *Drosophila* development but locates these within the more general context of developmental dynamics.

In Figure 7.1 is shown the metameric pattern of a wild-type *Drosophila* larva, with the distinctive pattern of denticle bands from anterior to posterior; and beneath it is the striking mirror-symmetrical pattern of a larva from a homozygous *bicaudal*-D mother. Somewhat less than half of the larval form is reflected in the ventral region, the segments being identified as a rudimen-

154

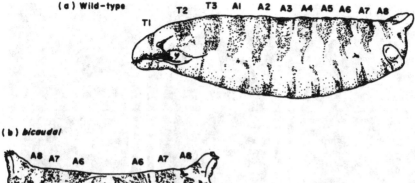

Figure 7.1. (a) Normal cuticular pattern in first instar larva; (b) mirror-symmetrical *bicaudal* phenotype.

tary abdominal 4 in the middle, with segments 5, 6, 7 and 8 and the terminal telson arranged in mirror-symmetrical array. Dorsally, segments 4 and 5 are missing, but the organisation of the form dorsal to ventral is perfectly coherent and without discontinuities. This is only one of a great variety of symmetric *bicaudal* mutant morphologies, which can have forms reflected about any A–P position between abdominal segment 5 and 8 – that is, with anywhere from two to five segments mirror symmetrised ventrally and with corresponding reductions dorsally. Thus there is no preferred mirror reflection plane, and the patterns are always coherently arranged from dorsal to ventral. Furthermore, asymmetrical mirror-duplicated morphologies are equally common with, say, four segments on one side and one segment on the other side of the reflection plane (Nüsslein-Volhard, 1977).

Weaker alleles of *bicaudal* result in a switch to a qualitatively different type of mutant morphology. Instead of mirror duplications, there are simply anterior deletions of different degrees of severity, from headless embryos with complete thoracic and abdominal segments, to embryos in which head and thoracic segments are absent (Figure 7.2). In the latter category are embryos in which rudimentary posterior spiracles are found at the anterior end of a complete abdominal region, which itself terminates posteriorly with well-

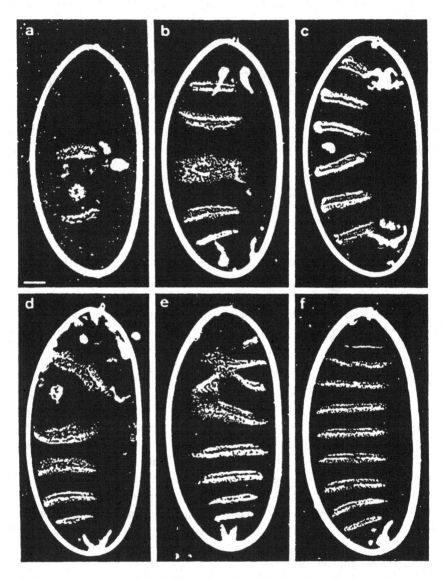

Figure 7.2. A range of phenotypes produced by different *bicaudal* alleles.

developed, normal spiracles. Next to the duplicated anterior spiracles are rudimentary mouth parts, characteristic of the anterior extremity of the embryo. The spatial juxtaposition of structures that are normally located at the opposite ends of the embryo is a very interesting observation. On the face of it, this seems to require a sharp discontinuity of state, since one expects the ends of the embryo to be maximally different from one another. However, it will emerge that this unexpected result is what is predicted by the explanation that will be advanced for the process that leads naturally to mirror duplication.

There are other even more striking discontinuities of pattern in mutant *bicaudal* phenotypes. It is possible for one side of an embryo to show a mirror duplication of some abdominal segments while the other is a headless phenotype with all eight abdominal segments and no mirror symmetry (Figure 7.2e). This means that the 'character' of a segment changes in the midline from being, say, abdominal 2 on the left of the embryo to abdominal 3 on the right, with a smooth transition between them. Even more remarkable are cases such as those observed in other insects, such as *Callosobruchas* (van der Meer, 1984) in which induced bicaudal phenotypes include embryos which are fully normal except for a longitudinal stripe in which a mirror duplication has occurred. How can such striking discontinuities arise? Clearly there are no well-defined compartments or subdivisions of the embryo that specify the positions of mirror reflection. Furthermore, there are substantial transformations of the components of overt morphology, such as the denticle band patterns. For example, in Figure 7.2b the bands are identified as two fused abdominal 6s in the centre, with abdominals 7 and 8 in mirror-symmetrical array. However, none of these segments can strictly be identified with normals; they are transformed in such a way that, if isolated, it would be difficult to name them. Similarly, in Figure 7.2e the mirror-duplicated abdominal 3 segment is much transformed in relation to a normal abdominal 3, and is identified as such simply by relative position, though there is nevertheless a significant change of denticle band structure in going from left to right across the midline. It is evident that mutations at a single locus can generate a considerable range of phenotypes, involving a broad range of transformed denticle band patterns. The same is true of wild-type genotypes under environmental perturbation. There is no fixed set of forms that is named by a genotype; rather, there is a transformation set with respect to short-range order such as denticle band patterns, and long-range order as in mirror-symmetrical phenotypes, together with all their intermediates. The dynamics of these transformations are clearly of the essence in developmental processes.

If the posterior part of a *Drosophila* embryo can be mirror reflected, it is to be expected that the same could happen to the anterior part so the embryo has two heads and no posterior. The gene whose mutant phenotypes have this form is called *dicephalic*. Both *bicaudal* and *dicephalic* are examples of maternal effect genes: it is the genotype of the mother that determines the phenotype of the embryo, resulting from the transfer of mRNA from the cells of the maternal ovary into the maturing oocyte. *Dicephalic* also shows the same spectrum of defects as *bicaudal,* strong alleles resulting in mirror-duplicated forms, both symmetric and asymmetric, while weaker alleles produce posterior deletions of different degrees of severity. Are there also genes active in the developing embryo that produce other mirror duplications and deletions? These do indeed occur, over characteristic spatial distances or wavelengths that are distinct from those that arise in the case of maternal effect genes.

Strong alleles of the mutant *Krüppel* result in mirror-symmetrical patterns, a typical phenotype being shown in Figure 7.3C: thoracic and anterior abdominal segments are missing, replaced by a partial mirror-image duplication of the remaining posterior abdomen which can include a set of rudimentary posterior spiracles near the head (Jäckle et al., 1988). Weaker alleles of *Krüppel* result in deletions of thoracic and anterior abdominal segments but no mirror symmetries. This is the same spectrum of transformations as *bicaudal,* but over a shorter spatial domain. A similar pattern to *Krüppel* is observed with the mutant *hunchback,* whose effects are centred on head and thoracic segments, these being either deleted by weaker alleles or replaced by mirror-image duplications of the anterior abdominal segments in cases of stronger alleles. *Hunchback* mutants also have a posterior domain of action, abdominal segments A7 and A8 being deleted. This is an important observation to which I shall return shortly. These two mutants belong to the category of segmentation gene called gap because of the characteristic pattern deletions, extending over four to seven segments.

The next group of segmentation genes is referred to as pair-rule, because of the typical pattern of deletions they produce and the wavelength over which they act, which is two segments. Mutant alleles of these genes result in, for example, deletions of every other segment: in *odd-skipped* there is an absence of segments T2, A1, A3, A5 and A7; while *even-skipped* has T1, T3, A2, A4, A6 and A8 missing. Other members of the set generate deletions on the same wavelength of two segments, but phase-shifted so that parts of adjacent segments are missing. Once again the now-familiar pattern of deletions in weak and mirror symmetries in strong alleles shows up. For example, strong *runt* alleles result in mirror-image duplications replacing the deleted domains, with the consequence that the eleven normal denticle belts are replaced by 6, 5 of which are mirror-imaged (Gergen and Wieschaus, 1985). The exact positions of the lines of mirror reflection vary from one embryo to the next, as in the gap and *bicaudal* examples of mirror-symmetrical patterns. Clearly we are dealing here with a generic property of

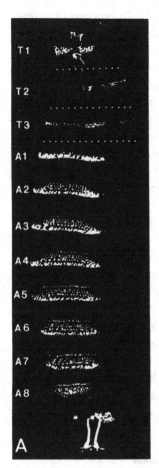

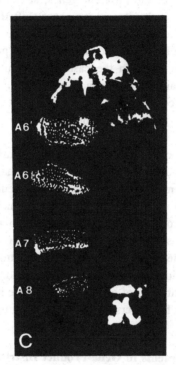

segments	Md	Mx	Lb	T1	T2	T3	A1	A2	A3	A4	A5	A6	A7	A8	A9
compartments	P	A	P A	P A	P A	P A	P A	P A	P A	P A	P A	P A	P A	P A	P A
parasegments		1	2	3	4	5	6	7	8	9	10	11	12	13	14

B

Figure 7.3. The typical phenotype of a strong allele of *Krüppel* (C) compared with the wild-type pattern (A). The range of *Krüppel* influence is shown in (B), in terms of segments affected and parasegments where the gene is expressed.

the patterning process involved in segmentation. A careful analysis of the *runt* pattern (Gergen and Wieschaus, 1985) showed that the mirror-image duplications are unlikely to be caused by cell death and regeneration but reveal a characteristic aspect of pattern transformation.

The last level of the hierarchy of segmentation genes, operating over wavelengths of a single segment, is in fact characterised by the property of mirror duplication. Members of this class are called segment polarity genes, because the typical effect of mutant alleles is a replacement of the posterior part of the denticle band by a mirror-symmetrical duplication (Nüsslein-Volhard and Wieschaus, 1980) though deletions also occur. The line of mirror reflection differs for each of the different genes, showing that the domains of gene influence within each segment are phase shifted with respect to one another. There is no evidence of any primary reference point within segments, these being generated by a periodic pattern with all positions (phases) as potential mirror-reflection lines. In the case of the mutant *patch,* for example, the duplicated domain involves structures of two adjacent segments so that, despite the presence of a normal number of denticle bands, there are twice the normal number of segment boundaries since each duplicated region includes this structure.

These observations show that there is a hierarchy of genetic effects on the segmentation process in *Drosophila* that is characterised by spatial patterns with distinct domains or wavelengths of influence. These extend from single segments up to nearly half of the embryo. Furthermore, within each of the four categories of segmentation gene, each operating over characteristic embryonic domains, there are qualitatively similar patterns of mutant phenotypes, weak alleles resulting in pattern deletions and strong alleles giving mirror duplications. It is necessary now to look beyond the phenotypes to the spatial distribution of gene products to see if there is a pattern that could provide a basis for understanding these intriguing observations.

Dynamic Transients in Gene-Product Distributions

The last two categories of segmentation genes, pair-rule and segment polarity, make it obvious that the segmentation process in *Drosophila* involves the occurrence of spatial periodicities of gene influence that are simple multiples of one another, with two-segment and one-segment wavelengths. Studies of the spatial pattern of gene transcription have amply confirmed the expectation from the mutant data that gene products should be found distributed in patterns of two-segment and one-segment wavelength for pair-rule and segment polarity genes, respectively. For example, Macdonald, Ingham and Struhl (1986) showed that the pair-rule genes *even-skipped* and *fushi tarazu* have well-defined two-segment periodicities of gene transcription at the late syncytial blastoderm stage (early in the fourteenth mitotic cycle), as shown in Figure 7.4E and K, with a relative phase-shift that correlates well, though not precisely, with the deletion zones of the mutants. *Engrailed,* classified as

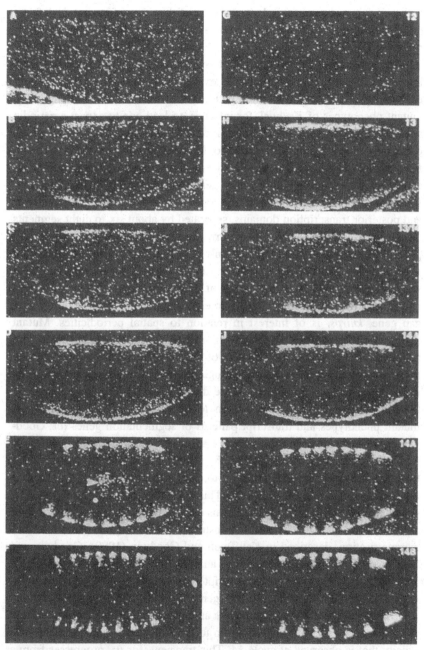

Figure 7.4. Transcript patterns of *even-skipped* (left) and *fushi tarazu* (right) revealed by tritiated thymidine labeled cDNA probes of *Drosophila* embryos at about 20-minute intervals starting in mitotic cycle 12.

both pair-rule and segment polarity because of the range of its mutant defects, has a one-segment pattern of transcripts at the gastrula stage (Weir and Kornberg, 1985).

What about the other categories of segmentation gene? In view of their longer wavelength effects, one would expect these to show correspondingly extended gene product distribution patterns. However, there is nothing in the mutant phenotypes of *Krüppel,* for instance, to lead one to expect a periodic spatial pattern of gene expression, since deletions and mirror symmetries occur only in one region of the embryo, located centrally. On the other hand, *hunchback* has two domains of influence, anterior and posterior, as remarked earlier, so we expect two domains of gene transcription. This has been confirmed: in early cycle 14 *hunchback* transcripts show well-defined anterior and posterior transcription domains, separated by about six to eight segments (Jäckle et al., 1986, 1988). But it turns out that the *Krüppel* transcription pattern is also spatially periodic: not only is there a broad central transcription domain centred on the region where mutant defects occur; there are also smaller anterior and posterior regions of transcription at early cycle 14, separated from the central region by six to eight segments. Finally, a third major gap gene, *knirps,* is of interest in relation to spatial periodicities. Mutant alleles of this gene result in a single six-segment deletion domain centred on abdominal 4, posterior to the domain of influence of *Krüppel.* Does its gene transcript pattern correspond to its single domain of influence; or is it, like *Krüppel,* spatially periodic, one of the domains of transcription being functionally silent? There is good reason to believe, from an analysis of the mutant phenotypes and transcript patterns of segmentation genes (cf. Goodwin and Kauffman, 1990) that *knirps* transcripts would show a periodic pattern with two bands, one centred on the posterior deletion zone, as expected, the other located anteriorly, some six to eight segments away, where there is no evidence of a mutant influence. This has since been confirmed (Tautz, personal communication). The analysis of this data, presented in Goodwin and Kauffman (1990), was used to construct a model of gene action in mirror symmetrisation, described in Kauffman and Goodwin (1990), Goodwin and Kauffman (1992). A summary of the analysis and the model now follows.

What is apparent from gene transcript data such as that presented by Macdonald et al. (1986) for the pair-rule genes *even-skipped* (*eve*) and *fushi tarazu* (*ftz*) (see Figure 7.4) is that there is an intriguing transient pattern, starting from mitotic cycle 13, and leading up to the expected double-segment periodicity that is observed at cycle 14. This transient, for *ftz,* progresses from a single broad band (H), through a doubly periodic (I) and then a roughly four-period distribution (J) before the seven bands are clearly defined (K). *Eve* goes through a similar sequence, and by late cycle 14 (F) there is the beginning of a 14-stripe (single-segment) pattern that becomes well defined later, in the gastrula stage. This interesting dynamic reflects the hierarchical organisation of the segmentation process, an initially uniform pattern of gene expression progressing to a well-defined spatial periodicity through a series of shorter wavelength transients. These are also observed in other segmentation genes. A detailed study of *Krüppel* transcripts by Knipple et al. (1985) and by Harding and Levine (1988) shows that these start during mitotic cycle 10–11 as a single broad band in the middle of the embryo, like the pair-rule pattern at cycle 12–13 (about 20 minutes later). This becomes a single well-defined central band during cycles 12 and 13, and in early cycle 14 the two additional transcript domains appear, located towards the poles. A fourth domain, posterior to the central band, appears towards the end of cycle 14 (early gastrulation) and by germ-band elongation *Kr* transcripts occur in all segments, but with a clear two-segment pattern of alternating intensities. It thus emerges that, despite the restricted domain of primary action of *Krüppel* as revealed in its mutant phenotypes, there is a pattern of spatial transformations in the transcript distributions that is very similar to those observed in the pair-rules, though starting earlier and finishing later. What we are seeing here is again a sequence of spatial patterns whose wavelengths get systematically shorter and the spatial frequency (the number of bands per unit of space) increases.

A similar sequence is observed initially for *hunchback* transcripts which have both maternal and zygotic contributions (Reinitz and Levine, 1990). These start with a uniform distribution that develops into a monotonic gradient with an anterior maximum by cycle 8, which is primarily of maternal origin. By cycle 12, a second peak appears posteriorly, and during cycle 14 the pattern develops three spatial periods, expressing the zygotic pattern. As yet there are no studies of later stages showing higher harmonics during gastrulation and germ-band extension, but it is anticipated that a pattern similar to that of *Krüppel* will emerge. These observations lead to the expectation that *knirps* transcripts will also show harmonic transients, but so far only the doubly periodic distribution has been confirmed. Another gap gene, *giant*, also has a doubly periodic expression pattern at midcycle 14 (Struhl et al., 1992).

What about the members of the first category of segmentation genes, like *bicaudal* and *dicephalic*? Do their gene transcripts also reveal dynamic transients that are much more complex than those anticipated from the mutant data? Neither of these maternal effect genes has been studied in detail, but a very thorough study has been made of *bicoid*, closely related to *bicaudal*

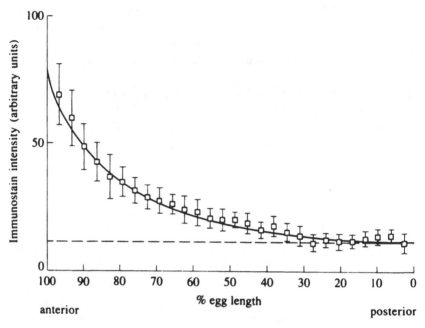

Figure 7.5. The spatial distribution of *bicoid* protein along the antero–posterior axis of an early *Drosophila* embryo.

(Driever and Nusslein-Volhard, 1988). mRNA transcripts of this gene, produced in the cells of the maternal ovary and transferred to the growing oocyte, are found localized at the anterior pole. Not until after fertilisation and egg deposition is *bicoid* protein produced from this mRNA, the protein then diffusing from this localised source to produce an exponential gradient that reaches background levels at about two-thirds of the distance to the posterior pole (Figure 7.5). This gradient of *bicoid* protein persists throughout the stages of early development up to the formation of the cellular blastoderm (cycle 14), after which it gradually decays. So there are no dynamic transients of the type observed with the other categories of segmentation genes! However, this is hardly surprising. Since the mRNA is of maternal origin, and the embryo contributes nothing, there is no developmental dynamic possible apart from translation of the mRNA. Only zygotically active genes can take part in the developmental transients. Evidently the maternal genes play a particular role in regulating early development, establishing certain essential initial and boundary conditions for the dynamic process that unfolds after fertilisation of the egg. A particular interpretation of this role will be given later in relation to a specific model of segmentation.

At this point it is useful to describe the effects of other maternal effect genes, of which there are estimated to be about 20 in all (reviewed in Govind and Steward, 1991). They fall into three categories, defined by their mutant phenotypes. *Bicaudal* and *bicoid* belong to the anterior class since their activity is required to produce anterior structures (head and thoracic segments). There is a complementary posterior class, of which *oskar* and *nanos* are typical. Mutations in these genes result in failure to form posterior (abdominal) structures; therefore their activity is required for normal development of these parts of the embryo. All the evidence points to a posterior localisation of maternally produced mRNA for this class of genes, and production of a gradient of protein after fertilisation that is stable until late cycle 14 (early gastrulation). Finally, there is a group of genes whose activity is required for normal development of the polar extremities of the embryo (the terminalia), of which *torso* is characteristic. Mutants of this gene result in an absence of the acron (anterior structure) and the telson (posterior structure).

The evidence points conclusively to a graded distribution of gene products for members of this class, at both poles. This is of interest in relation to the observations suggesting that spatial periodicities and harmonics are of basic significance in the organization of the body pattern. If the embryo is organised in a monotonic manner from anterior to posterior, as by gradients, then it is expected that the extremities would be most different from one another. If, however, spatially periodic patterns are fundamental to the pattern-forming process, then the two ends of an embryo could be similar to one another, just as the two ends of a segment are more similar to one another than either is to the middle if an underlying pattern repeats itself in every segment. There is strong evidence for the latter in terms of the segmental periodicities of gene transcripts such as *engrailed* at the germ-band stage, when segments are taking definitive form. Now we see some evidence for such periodic organisation over the whole embryo. This begins to explain why, in *bicoid* mutants, it is possible to have rudimentary posterior spiracles at the anterior end of an abdominal region, adjacent to mouth parts (cf. Goodwin and Kauffman 1992).

It is interesting to recall at this point Bateson's observation that the members of a meristic series such as segments are in certain respects like a series of individual organisms. This is in fact made explicit in an interesting article by Buss and Dick (1992), who point to the parallels between segmentation gene domains in *Drosophila* and asexual reproduction patterns in annelids, which include segment skipping, mirror-image duplication and ectopic duplication of structures. Segments are the units of asexual reproduction in these worms, as they are of regeneration (Berrill, 1952). The principles involved in the relationships of parts of an organism are then similar to those that relate the various parts of different organisms, so that homologies exist both within and between species. This gives us a powerful unifying principle in the study of biological form and transformation, allowing us to make phylogenetic (taxonomic) deductions from ontogenetic observations. However, these principles have not yet fully emerged, so let us return to the *Drosophila* analysis to complete the story.

The absence of transient patterns in maternal gene products is, as observed earlier in this discussion, a natural consequence of the fact that embryonic genes are not involved in transcript production. However, there is a very interesting gene called *caudal* with a spectrum of mutant phenotypes including posterior deletions of *oskar* type, which has both maternal and zygotic transcripts. Does it conform to the expected pattern of harmonics in gene-product distributions? Macdonald and Struhl (1986) observed that *caudal* (*cad*) protein is initially at a uniformly low level throughout the egg, but accumulates rapidly during cell cycles 7–9 in the posterior half of the embryo to produce a sigmoid pattern, low anteriorly and rising to a flat plateau posteriorly. During cycle 14 the central gradient steepens and there is a decrease at the posterior pole, resulting in a band in the midposterior region of the embryo. During germ-band extension a double-segment pattern of *cad* protein is superimposed on this pattern. Mutant phenotypes add to this picture. Some larvae show deletions in abdominal segments 4 and 8, hinting at a four-segment influence, while others have partial deletions of even-numbered abdominal segments, showing a pair-rule pattern. In general, *caudal* phenotypes involve deletion of A8 and posterior terminalia, which are replaced by structures resembling mouth hooks. This illustrates once again the ease with which the poles transform one into the other.

The molecular evidence shows that underlying the hierarchical phenotypic effects of the different categories of segmentation genes lies a dynamic pattern of gene-product distributions with characteristic harmonic transients. Those genes with short-wavelength influences, such as segment polarity and pair-rule, start late and pass through the transients rapidly, whereas the genes with longer wavelength effects start their transients earlier and progress more slowly to double- or single-segment distributions. The strictly maternal gene transcripts are under translation control only, and there is no evidence of transients. The overall result of this complex dynamic pattern of spatial gene-product distribution is that the embryo has available to it a hierarchy of wavelengths of gene expression that influences spatial patterning on progressively finer spatial scales. The whole system has a characteristic space–time order that involves spatial periodicities as a major aspect of the pattern generator. At the moment no particular mechanism need be assumed for this process, and the analysis that follows is independent of how the periodic spatial patterns of segmentation gene products are generated. The majority view amongst biologists is that these patterns arise from combinatorial inter-actions between members of the different segmentation gene hierarchies, as described in models of the type presented by Meinhardt (1986), Akam (1987), Ingham (1988), Struhl et al. (1992) and Lawrence (1992). The detailed mod-elling of the combined effects of maternal and gap genes on *eve* expression by Reinitz and Sharp (1995) is the most comprehensive and revealing anal-ysis yet carried out on the dynamics of Eve protein stripe formation. Their gene circuit method shows how the complexity of regulatory interactions can be resolved to provide a consistent picture of the way in which the inductive effects of *bcd,* and the repressive influences of the gap genes, together result in a periodic pattern of Eve expression, with a transient during cycle 14 that closely matches experimental data. This modelling procedure promises to reveal some essential aspects of gene regulatory processes in terms of inter-action kinetics and diffusion wherever the data on gene expression in wild-type embryos is sufficiently detailed to guide the modelling procedure through parameter modification and simulated annealing to a solution. The next step in our enquiry is to see how spatial periodicities result naturally in the characteristic deletion to mirror-duplication pattern for weaker to stronger alleles in the segmentation process.

Spatial Periodicities and Mirror Duplications

Since the same phenomenon arises on different wavelengths, it does not mat-ter which example is used to relate spatial periodicities to mirror duplications. Furthermore, the model that will now be constructed to explain the charac-teristic spectrum of mutant phenotypes, from deletions to mirror symmetries, is independent of *how* the patterns of segmentation gene products are gen-erated. It requires only that the genes within any segmentation category have domains of expression that are systematically phase shifted relative to one another, as observed, together with another property of the map from gene products to expressed patterns that will shortly be described. For purposes of clarity, consider the two-segment periodicities that are observed at cycle 14 in the three major pair-rule gene transcripts, *hairy, eve* and *runt,* or the same pattern in the corresponding translation products, the pair-rule proteins (Fig-ure 7.6). Each of the different gene products has a characteristic phase rela-tion to the others, the maxima and minima being shifted relative to one another so that they span the two-segment cycle in a well-defined order. To

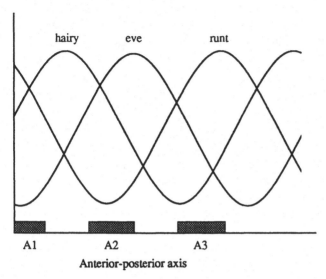

Figure 7.6. The approximate phase relations of the periodic patterns of the three major pair-rule gene products, *hairy*, *eve* and *runt*.

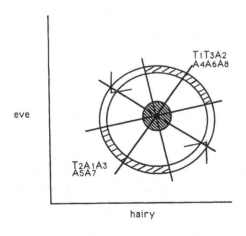

Figure 7.7. A plot of *eve* against *hairy* over one spatial cycle, corresponding to a pair of segments, with the arrow defining the antero-posterior direction along the body axis. The approximate positions of the denticle bands are indicated by the shaded domains, with even and odd abdominal segments (and conversely for thoracic segment numbers) located as shown. The shaded central region is the null disc where no pattern elements are defined.

demonstrate the principle involved, consider first only two genes, say *hairy* and *even-skipped*. These have a relative phase shift of about 110–120°, as shown in the schematic periodicities of Figure 7.6, which represent the concentration of protein products in relation to position in the embryo. The curves are idealized, but as the argument that follows is essentially topological, this does not affect the deductions. The position of the denticle bands, using A_1, A_2, A_3 for illustration, are shown in relation to the pair-rule gene product periodicities, there being two of these per cycle.

Now plot *eve* against *hairy* over one cycle, as shown in Figure 7.7. The result is a closed circle in which spatial position in the embryo (the abscissa or x-axis in Figure 7.6) is represented by position on the circle following the arrow for the direction anterior to posterior. The position of the denticle bands, of which there are two per pair-rule gene cycle, are as shown. Since the cycle repeats along the embryo, mapping two segments per pair-rule cycle, we can assign segment identities to the denticle bands in groups as shown, T_1, T_3 and the even-numbered abdominal bands being mapped at one position on the repeating cycle, while T_2, and the odd-numbered abdominal segments are identified with the other. The pair-rule genes do not, of course, themselves specify differences between the segments, which result from the combined influence of all segmentation genes together with the homoeotics.

Reading Gene Products as Ratios

It is necessary now to consider what general type of functional relationship there is between gene products and the patterns of cell differentiation for which they code. It is often assumed that genes specify pattern elements by a binary combinatorial code that depends upon the presence or absence of gene products in different combinations. Effectively, threshold levels can be specified above which a gene product contributes to pattern specification and below which it does not. Suppose that there are four segmentation genes, A, B, C and D, involved in specifying pattern at one level of the segmentation gene hierarchy, phase shifted relative to one another as shown in Figure 7.8a. Take the threshold value to be the middle of the amplitude variation, and assign a + to values above, and a − to values below this reference value. Now construct a circular plot of these variables as was done for *eve* and *hairy,* showing the points where they change from above to below threshold and coding the eight sectors according to the binary combinations, as shown in Figure 7.8b. Suppose next that there is a mutation in gene A so that the concentration of its product is everywhere below threshold. The resultant combinatorial code words will then be as shown in Figure 7.8c. The result is not a mirror-symmetrical pattern but one in which two sectors have meaningless or illegal code words. Therefore a simple binary combinatorial code for the relationship between gene product and pattern is not consistent with the observations on mirror-symmetrical mutants. It is therefore necessary to consider a different mapping.

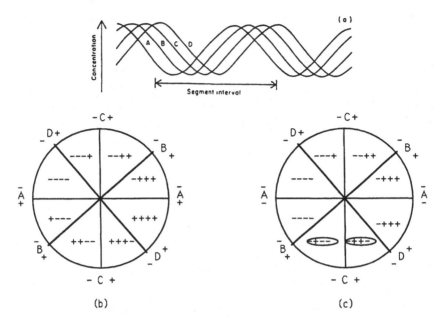

Figure 7.8. The predictions of a binary code for gene action on pattern formation. Four genes whose periodic spatial patterns are phase shifted as in (a) generate the codings shown in (b) over one cycle, assuming that the midpoint of their amplitude is the threshold that distinguishes between + and −. A mutation in gene A fails to produce a mirror-symmetrical pattern (c)

Some observations of Gergen and Wieschaus (1985) and Gergen, Coulter and Wieschaus (1986) are particularly relevant to this question. Studies of embryos with different numbers of wild-type *runt* alleles showed that when *runt*[+] copy number is decreased, the odd-numbered denticle bands are deleted, giving a phenotype in the general *odd-skipped* category. However, when the copy number of *runt*[+] is increased, a complementary antirunt phenotype arises in which deletions occur in the even-numbered denticle bands. So overproduction of *runt* results in a phenotype similar to that produced by loss of function at other pair-rule loci such as *even-skipped* or *paired;* that is, too much *runt* is like too little *eve* or *prd*. Too little *runt*, on the other hand, results in an *odd-skipped* phenotype. Genes evidently do not affect pattern by a cumulative linear measure along a single scale of influence, such as progressive effects on the same pattern elements, but interact in some more complex manner.

Another striking observation (Coulter and Wieschaus, 1988), relates to the consequence of double mutants of *even-skipped* and *odd-skipped.* From the deletion effects of these two genes taken individually, the double mutant would be expected to lack both even- and odd-numbered denticle bands. Instead, the double mutant has eight partial denticle bands each of which is a small mirror duplication of the normal patterns of abdominal segments 1–8. So the elimination of a second gene can result in the recovery of pattern elements deleted by the absence of a first gene. These and the *runt* results suggest that genes may code for pattern elements by the ratios of their concentrations, or some other function that is not a linear measure of product concentrations.

We can now return to Figure 7.7 and complete the description of the relationship between the spatial distributions of pair-rule gene products and the specification of pattern. Let us assume that the effect of gene products on pathways of cell differentiation is determined by the slope of the line passing through the pair of gene-product concentrations (x,y) as measured from the centre of the circle. Each radius from this centre then defines a set of equivalent gene pairs, with slope given by tan $\vartheta = y/x$, the ratio of the concentrations. Here ϑ is the angle or phase of the radius relative to a reference line, taken to be through the centre of the circle and parallel to the x-axis. The different angles defined by the specific case of ratios of *eve* and *hairy* in Figure 7.7 map into segmental patterns as shown, each full cycle for these pair-rule genes defining pattern elements over a pair of segments. Of course the detailed character of the denticle band pattern involves influences from other pair-rule genes as well as other segmentation genes and the homeotics. Figure 7.7 is a projection of a much more complex mapping onto the subspace of *eve* and *hairy* concentrations.

The point of intersection of all the radii at the centre of the circle has all angles and so defines a singularity of the mapping, a phaseless point. The pattern specification process has a limited degree of resolution relative to ratios, so this phaseless point will effectively be a null domain of finite extent – the disc of Figure 7.7 – within which no specific pattern elements can be discriminated. Similarly, there will be discrete sectors defined by ranges of phase (ratio) within which no discriminations are made, all cells within such a range following the same pathway of differentiation in terms of the contributions of these two genes. So the whole space, called by Winfree (1980) tissue specificity space (TSS), is quantized. In Kauffman and Goodwin (1990) these quantized sectors were identified metaphorically with the colour spectrum, and the sectored cycle was described as a colour wheel. This language reflects a topologically similar analysis carried out by Winfree (1984) on the periodic temporal organisation of organisms, particularly biological clocks, in which the concept of the isochron was introduced to describe states that map into points of equal time in the dynamic space of biological oscillators. The description given here of the dynamics of spatial organisation in organisms by periodicities in space rather than in time has deep qualitative similarities to that presented by Winfree, and some of these properties will be developed further in a later section. Clearly an objective is to join the two in a unified analysis of the space–time organisation of developing and behaving organisms as dynamic systems of a particular kind, with generic properties that come from basic topological features. We are now in a position to see how these relate to deletions and mirror symmetries.

Reading Epigenetic Maps on Four Colour Wheels

Referring to Figure 7.9a, loss of function mutations in *eve* correspond to displacement of the circle towards the *hairy* axis and a flattening that describes a reduction in amplitude of the *eve* product. When this displaced curve meets the null disc, as shown by the elliptical curve, pattern deletions will occur in the even-numbered denticle bands. The odd-numbered bands will continue to be generated, since the dotted curve intersects the radial lines defining the mapping for those pattern elements outside the null disc. Thus the *even-skipped* phenotype results.

A similar analysis can be carried out for *hairy*. The phenotype of a typical weak mutant is characterised by loss of the anterior part of each even-numbered abdominal segment and the posterior part of each odd-numbered segment, with corresponding deletions to the thoracic segments. The effect of such a mutation is a distortion of the colour wheel towards the *eve* axis, resulting from decreased levels of *hairy* products. When this distorted curve intersects the null disc (Figure 7.9b) there will be a loss of anterior parts of even-numbered segments and posterior parts of odd-numbered segments, as observed. In stronger hypomorphic alleles, the curve will cross the null disc and come to lie entirely to the left of it (Figure 7.9b), so that T_1, T_3 and the even-numbered denticle bands entirely disappear. Furthermore, the curve will now intersect each of the lines in two points. (The lines of equal gene ratios extend throughout the quadrant.) Following the curve around and identifying the pattern specified, it is evident that one part of the curve specifies the odd-numbered abdominal segments in the normal spatial order while the other part, intersecting the same lines in the opposite direction, specifies the same pattern elements in mirror-symmetrical order (and similarly for T_2). So strong *hairy* alleles should result in mirror-symmetrical odd-numbered abdominal denticle bands, which they do (Ingham et al., 1985). However, this phenotype has no naked cuticle between the mirror-symmetrical denticle bands, which

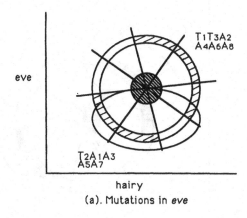

eve

(a). Mutations in *eve*

Figure 7.9. (a) The predicted result for loss-of-function mutations in *eve* is shown by the ellipse that intersects the null disc, with corresponding loss of denticle bands T_1, T_3, A_2, A_4, A_6 and A_8. (b) Mutations in *hairy* giving either pattern deletions or mirror-symmetrical patterns depending on the strength of the allele. (The spokes of the colour wheel extend throughout the quadrant.)

eve

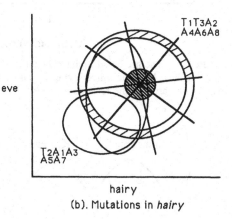

(b). Mutations in *hairy*

constitute a continuous mass of setae. So the distorted curve in Figure 7.9b is placed in the lower quadrant. This is actually what is expected if we take account of the pattern of regulatory interactions believed to exist between the pair-rule genes; *hairy* represses *runt* which represses *eve* (Carroll and Vavra, 1989). Therefore decreased levels of *hairy* product will result in elevated *runt*, hence repressed *eve*. So the observed strong *hairy* phenotype is what is expected from the model.

This procedure can now be used to explain the effects of *runt* alleles. *Runt* is plotted against *hairy* to give the relationships shown in Figure 7.10, with the denticle band patterns located on the circle as indicated. Too much *runt* means moving the curve towards the right and down, since *runt* reciprocally represses *hairy*. When this intersects the null disc, deletions of the even-

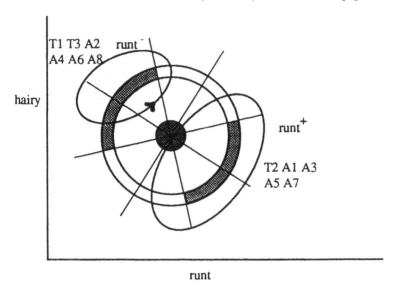

Mutations in runt.

Figure 7.10. The predicted mutant phenotypes for too much or too little *runt* product, showing how both the mirror-symmetrical odd-skipped pattern and the even-skipped pattern can arise in the respective cases.

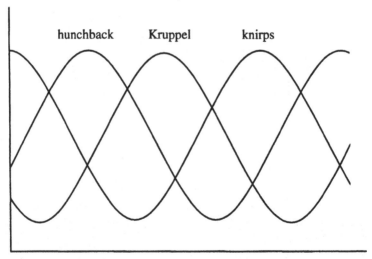

Anterior-posterior axis

Figure 7.11. The approximate phase relations of the periodic distributions of the major gap gene products.

numbered denticle bands occur, giving the even-skipped phenotype. Conversely, too little *runt* corresponds to displacement of the curve to the left and up (*hairy* is derepressed by reduced *runt*), which results in odd-numbered denticle band deletions, the complementary effect, as observed. But it is now possible to understand also the mirror-symmetrical phenotype of strong *runt* alleles. These carry the curve up and to the left of the null disc, as shown. Following this curve around in the direction of the arrow (the A–P direction) gives a mirror-symmetrical map: the rising part of the curve passes across the sectors that map the anterior part of the even-numbered denticle bands and the posterior part of the adjacent (anterior) segment, as described for *runt*[YE96] (Gergen and Wieschaus, 1985). So we get naturally the full range of mutant phenotypes described for *runt* which at first sight seemed so strange. The phenomena are generic consequences of spatial periodicities in gene products and using ratios of gene product concentrations to determine patterns of cell differentiation.

The model also provides predictions. Stronger *runt*[+] embryos in which the curve is displaced further to the right should result in a mirror-symmetrised phenotype. In fact, varying the concentration of any segmentation gene product from low to high should produce the full range of phenotypes from one mirror-symmetric pattern to deletions to normal and then through the complementary deletion phenotype to the complementary mirror-symmetrised form. Such experiments can now be done not just with mutants, but more directly by injecting antisense RNA to reduce gene product or injecting mRNA to increase it. And various gene constructs will do the same. So this constitutes an experimental test that would carry further the analysis of the proposition that genes exert their influence via ratios in the particular manner described here.

The three major gap genes, *hunchback, Krüppel* and *knirps,* have phase-shifted distributions of gene products during cycle 14 that cover the segmentation domain in the manner shown in Figure 7.11. Plotting *hb* against *Kr* and locating the segments on the closed curve that now represents part of the antero–posterior axis of the embryo gives us Figure 7.12a. Strong mutations in *hb* result in the elliptical curve shown, with deletions occurring in the head and thoracic segments while mirror-symmetrical patterns involve the anterior abdominal segments, as observed. The equivalent plot of *knirps* and *Krüppel* is given in Figure 7.12b, which shows mutants in *Kr* deleting the thoracic and abdominal segments (T1 to A5 – see Figure 7.3b), while stronger ones result only in A_8 and A_7 and a mirror-symmetrised A_6. These two-dimensional projections need to be put into a three-dimensional context to identify the detailed characteristics of the mutant phenotypes involving

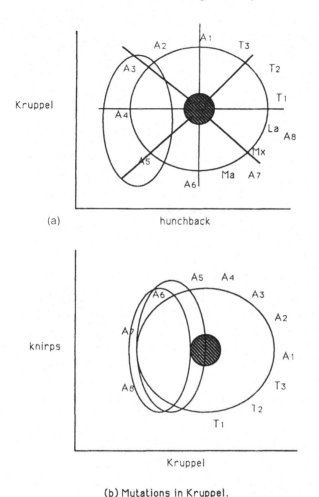

(a)   hunchback

(b) Mutations in Kruppel.

Figure 7.12. The predicted phenotypes of hypomorphic mutations in *hunchback* (a) and in *Krüppel* (b), showing the expected patterns of segment deletion and mirror symmetry. Ma = mandibular, Mx = maxillary, La = labial segments.

interactions among all three of the primary gap genes. This should also be supplemented by the further contributory effects of the other members of this group such as *giant* and *tailless*, and possibly *unpaired* and *hopscotch* (Akam, 1987).

The primary gap genes, *hb, Kr* and *kni*, are intimately involved in the process whereby the periodic pattern of pair-rule genes products is generated. This emerges from the effects on pair-rule gene activities in mutants of both the maternal genes, which affect gap gene activity patterns, and mutations in the gap genes themselves (Carroll and Scott, 1986; Carroll and Vavra, 1989; Gaul and Jäckle, 1989). How a periodic pattern of seven stripes arises within a domain with a distribution of primary gap genes of the type shown in Figure 7.11 remains a puzzle that has led Gaul and Jäckle (1989) to propose that "gap gene concentration levels, and ratios between concentration levels, have to be taken into account." It is the very complexity of gene interactions that has overwhelmed any simple account of the causal relationships between the activities of the segmentation genes and the pattern of stripes that arises. However, this complexity, recognised long ago as the prevalence of epistatic interactions in developmental processes, has the characteristics of a system that works as a network within which robust states of order can arise. The general features of such a genetic network and the way it mediates between development and evolution will be considered shortly. For the moment, I return to the analysis of the topological properties that underlie mirror symmetrisation in subsystems of the network.

Applying the colour wheel principles to maternal genes requires a modification of detail, but not of topological principle. The protein product of the *bicoid* gene is distributed in an exponential gradient with maximum at the anterior pole of the embryo (Driever and Nüsslein-Volhard, 1988), and presumably *bicaudal* is similar. It is assumed that other maternals have similar spatial patterns though with maxima at other positions. *Nanos* and *oskar*, for example, take their maxima at the posterior pole. Plotting *bicaudal* against *oskar*, for instance, gives a curve of the general type shown in Figure 7.13. The origin of coordinates is taken to be near the mean value of the range of the two variables. Quantizing the domain to specify the ranges of resolution of the variables results in a fan rather than a colour wheel. What happens now in a *bicaudal* deficiency mutant? The level of *bic* will clearly fall; but now *oskar* product starts to rise at the head end, where it is normally repressed, and so the curve distorts towards the origin as shown in Figure 7.14. With greater reduction in *bic* the distortion increases and the curve passes through the null zone giving anterior deletions. Further distortion results in a loop that cuts the posterior sections twice, giving a mirror-imaged bicaudal embryo. The mirror-imaged domain can be symmetric or asymmetric, depending on the extent of distortion of the curve. The curious phenomenon of duplicated spiracles at the anterior end of a headless embryo, with mouth parts adjacent to the duplicated spiracles, also finds an explanation in Figure 7.14. The curve that just touches the lower part of the null disc describes an embryo lacking anterior parts (say head and thorax missing). However, the curve cuts the spiracle domain (shown as a line) and, next to that, the mouth

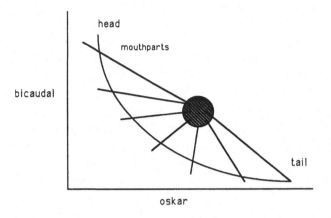

Figure 7.13. Pattern specification by ratios of maternal genes *bicaudal* and *oskar*, measured from a central origin (the null disc).

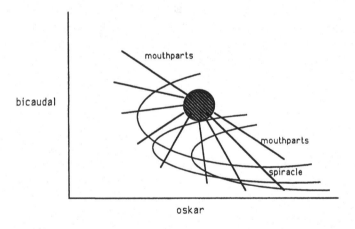

Figure 7.14. The predicted phenotypes for mutations in *bicaudal*, showing how deletions and mirror symmetries occur.

part domain (also shown as a line, though both are really sectors). The latter is the continuation of the normal mouth part domain on the opposite side of the null disc. This interpretation of the pattern of such embryos is possible only if the origin of coordinates for gene ratios is in a position such as that shown in Figure 7.14.

Turning to the other striking categories of mutant morphology mentioned earlier, such as those with mirror duplications on one side but deletions on the other, it is now evident how these come about in terms of the model proposed. The distance between curves describing deletions and mirror symmetries (cf. Figures 7.9b, 7.14) is small in terms of concentration differences of gene products, so neighbouring regions of the embryo can undergo such transitions of morphology. Similarly, phenotypes with a longitudinal strip of mirror-symmetrical patterning can be understood as regions where the curve in tissue specificity space gets distorted locally, passing across the null zone and into the mirror-symmetry region by a perturbation. So the topological properties of these mappings with singularities make the range and mixture of phenotypes observed much easier to understand than interpretations of gene products read as concentrations.

This discussion has shown how the general principles of the analysis extend to the longest wavelength category of segmentation genes, the maternals. Going in the direction of shorter wavelengths, from pair-rules to the segment polarity genes, clearly presents no difficulties since these are treated in the same way as pair-rules, but on a wavelength of one segment. Mirror symmetry is the defining character of mutations in these genes, polarity reversal of part of the segmental pattern being their distinctive phenotypic character.

The overall result of this model is a set of four 'colour wheels' that define four multidimensional mappings of spatial distributions of gene products, each with a characteristic wavelength, onto the final spatial pattern of differented cells in the *Drosophila* larva. Each colour wheel (or fan), from maternals to segment polarity genes, specifies pattern on finer and finer spatial grids, the whole hierarchy constituting a coherently organised, nested dynamical set. This coherence comes from interactions between and within members of the different groups of genes that affect their spatial and temporal patterns of activity, as previously described. Thus segmentation genes not only influence pathways of cell differentiation, resulting in spatial patterns; they also modulate one another's activity, resulting in coordinated spatial and temporal patterns.

Epigenesis and Teratology

I have used the conventional terms 'genotype' and 'phenotype' to identify the spaces between which the mapping, involving ratios of gene products, has been described, represented by morphological states on the colour wheels corresponding to the different regions of gene-product state space. This must now be recognized as a simplification. The mapping is in fact from epigenetic space, not gene product state space, to morphological space (Winfree's tissue-specificity space). Epigenetic space (see Goodwin, 1963) involves many variables other than gene products, including all the primary regulatory metabolities and ions such as cAMP, IP_3, diacylglycerol (DAG), Ca^{2+}, H^+, mechanical state of the cytoskeleton, and a host of other variables. Gene products constitute only one set of variables of the process, the ones that become most evident in genetic analysis. Hence the stress in the present chapter, which has focussed on a generic property revealed most clearly by genetic studies. However, the existence of a complementary set of variables that define the epigenetic system is evident from the perturbation studies that generate the range of teratologies whose analysis constitutes classical embryology, and particularly the phenomenon of so-called phenocopies. Perturbation of development by mutation can be mimicked by physical or chemical stimuli applied to genetically normal embryos, diverting the epigenetic system into pathways other than the normal (Goldschmidt, 1945; Waddington 1957; Ho et al., 1987; Matheson, 1991). A striking example of this is shown in Figure 7.15, a bicaudal form resulting from ether perturbation between 1 and 1½ hours after egg deposition of a *Drosophila* embryo. Another example is provided by the ether-perturbed *Drosophila* embryo in Figure 7.16, which shows a spiral pattern of *engrailed* gene product of a type not to my knowledge reported for mutant studies. These results were obtained by Lynda Micklewright working in my laboratory.

Another very important technique of investigation into the variables of the

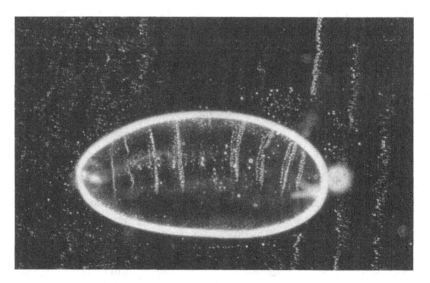

Figure 7.15. A phenocopy of a *bicaudal* mutant produced by ether shock between 1 and 1½ hours after egg deposition.

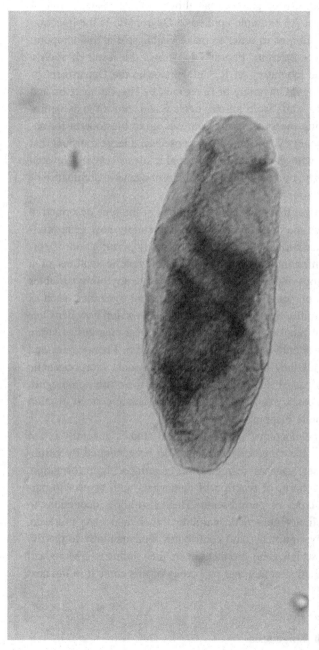

Figure 7.16. An unusual spiral pattern of *engrailed* gene product resulting from ether perturbation of an early embryo.

This plate is available for download in colour from www.cambridge.org/9780521207430

epigenetic system is the use of probes that allow a direct visualisation of developmental dynamics. An example applied to *Drosophila* is the injection of aequorin into early embryos in order to make visible spatial and temporal changes in cytosolic free calcium. Figure 7.17 shows the result of such a study, carried out in the laboratory of R. Cuthbertson in the Department of Anatomy and Cell Biology, University of Liverpool, by Ho, Cuthbertson and Goodwin. The recording starts from mitotic cycle 6 and shows the dynamics of calcium change during the synchronous mitoses up to blastoderm formation, when there is a longer cycle time superimposed on a large transient that is probably associated with gastrulation. A spatial analysis reveals elevated calcium levels posteriorly, corresponding to the morphogenetic dominance of the posterior pole in *Drosophila*.

These studies emphasize that epigenetic mappings of the type described in this chapter do not have the objective of identifying simply how gene products exert their influence on development, but how *any* perturbation of epigenesis is likely to enter into the determination of generic features of a mapping. The conclusions about gene products extend to any other variables that exert an influence on some aspect of developmental dynamics, such as change of membrane fluidity by organic solvents; change of ion flow densities by ionophores, channel blockers or magnetic fields; local changes of strain of the cytoskeleton by alteration of cell shape; and so on. Phenocopies and the mutant morphologies they mimic result from equivalent morphogenetic processes even though causal details differ. They are therefore homologues of one another. It is necessary to keep this broader dynamic context in mind in any study of epigenetic mappings.

The importance of phenocopies and teratological studies generally is not simply that they reveal the range of forms that can be generated by particular epigenetic processes, such as segmentation or tetrapod limb formation, which define a kind of norm of reaction of epigenesis with respect to particular characters and types of perturbation. They also allow deductions to be made about the primary epigenetic variables other than gene products, whose action is equally significant and contributes fundamentally to the dynamic context within which gene products exert their influence. More will be said about the detailed structure and properties of this context in the next chapter.

Generic Properties of Epigenesis

The four colour wheels model is not a description of the morphogenetic field that generates the pattern elements in *Drosophila*. It does no more than iden-

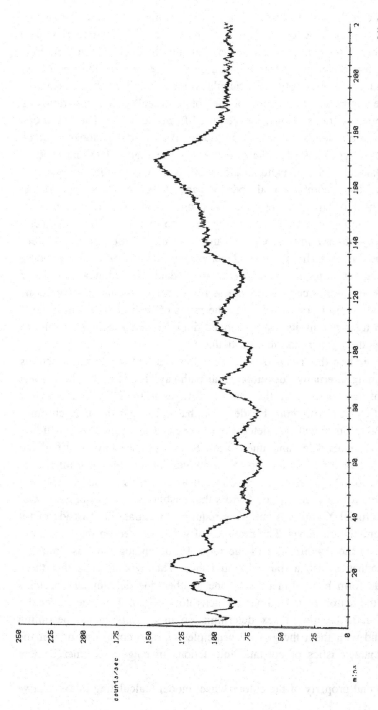

Figure 7.17. Periodic changes in cytosolic free calcium in a *Drosophila* embryo observed after aequorin injection and measurement of light emission with a sensitive photomultiplier. The recording starts at cycle 6 and each small cycle corresponds to a mitotic cycle. The first elevated phase appears to be due to nuclear migration, and the second to gastrulation.

tify a topological property of the epigenetic mapping that could account for what appears to be a generic feature underlying the mirror-symmetrical patterns that occur on a discrete set of wavelengths in segmentation gene mutants. Two aspects of this mapping stand out: (1) genes exert their influence through ratios; (2) these ratios are calculated relative to reference concentrations of gene products (the centre of the colour wheel) rather than ratios of absolute concentrations of product measured from the origin. The first property points to a redundancy in gene influence: the two-dimensional space of a pair of gene products (e.g., *hairy* and *eve* – see Figure 7.7) maps into a one-dimensional space of patterns along the A–P axis of the embryo. We saw that a binary combinatorial code of gene influence of the type that is often assumed is unable to reproduce the mirror-symmetrical mutants observed, so that a more complex mapping needs to be used. Ratios have been suggested by observations of Coulter and Wieshaus (1988), and they were also proposed in an earlier theoretical analysis by Russell (1985). They appear to be the simplest combinations that explain the data. The redundancy implied gives to gene action a certain robustness in that gene product concentrations need not be precisely regulated; only the ratios of their concentrations need fall within tolerance limits (the quantized domains of the colour wheels) to influence pattern generation in a particular way.

The recognition that ratios of gene activities can be the relevant functions in determining alternative developmental pathways is not new. The earliest example of this comes from the work of Bridges in the early 1920s on sex determination in *Drosophila*. He deduced that it is the ratio of X chromosomes to autosomes (A) that determines the sex of an individual (including metafemales, metamales and intersexes). Recent studies have clarified the nature of the factors contributing to the numerator and the denominator of the X:A ratio. The products of genes such as *sisterless-a* and *-b* (*sis*) and *daughterless* act as transcription factors that enable the activity of *Sex-lethal* (*Sxl*), the crucial X-chromosome gene required to initiate the cascade of female differentiation genes. The products of genes located on the autosomes contribute to the denominator of the ratio. Prime among these is *deadpan*, whose product is also a transcription factor. The hypothesis is that these factors can form heterodimers with one another, the denominator proteins blocking the action of the numerator activators of *Sxl*. Thus the molecular details of ratio control of sex determination in *Drosophila* are becoming clearer, and with them the general principles of ratio control as one of the generic characteristics of epistatic interactions in epigenesis emerge more clearly.

The second property of the colour wheel model (calculating ratios relative to a reference value) suggests that gene products can alter the nature of their effects according to their concentration (e.g., behaving formally as an activator over one concentration range and as an inhibitor over another). This is actually a general property of control systems involving the action of two complementary signals, as illustrated by the following recent example. It has been shown by Hemmati-Brivanlou, Kelly and Melton (1994) that in *Xenopus* embryos activin, a protein belonging to the TGFβ (Transforming Growth Factor) family, is involved in mesoderm induction when present above a critical concentration, but neutralization of competent tissue occurs when the activin signal is effectively below a threshold (e.g., blocked by follistatin, an antagonist). What is important for the net effect of these two gene products is their ratio, [activin]/[follistatin], and there is a switch from one effect on development (mesoderm induction) to another (neural induction) when this ratio decreases below a critical value. So long as the ratio is constant, the effect is constant. This kind of complementary action of different signals is therefore also likely to be a widespread characteristic of epigenetic control systems, so that the use of ratios with thresholds to describe epigenetic maps may emerge as a generic property.

A major limitation of the four colour wheel model is that it gives a static picture of a highly dynamic process. It is clear from the data on gene expression described at the beginning of this chapter that patterns of gene products are in continuous transformation during embryogenesis, and static mappings of the type proposed are only snapshots of an unfolding process. Redundancy does make certain dynamic changes irrelevant to the mapping (e.g., gene products changing proportionately such that their ratios are unaltered) and there may be much higher levels of redundancy that result in other invariants that contribute to epigenetic robustness. However, an understanding of the generic properties that underlie biological forms requires more than the characteristics of the mapping from genotype to phenotype, which is only a part of the story.

A more dynamic picture is obtained if we extend the *Drosophila* study in the direction of a comparative analysis with other, more primitive, insect species. Investigations of grasshopper, beetle and moth (Patel, 1994) reveal certain similarities of gene expression but also considerable differences, particularly between short- and long-germ species (French, 1993). It is difficult to uncover basic patterns across these species that identify the common elements of a segment generator, but so far the evidence points to the pair-rule and segment polarity genes as fundamental, conserved elements. Hunding and Engelhardt (1995) have presented an interesting case for these as basic contributors to the excitable kinetics underlying segment formation. In short-germ insect species, which are the more primitive forms, segments are generated successively in a posterior growth zone by a wave that appears to propagate with the characteristics of a Beloussov-Zhabotinsky reaction: concentrations of substances rise and fall periodically as may be the case for *hairy* and *engrailed* products during successive segment formation in *Tribolium castaneum* embryos. The segmentation gene products could be interacting to produce cooperative associations or multimers, giving the high-order nonlinearities required to generate the kinetic instability underlying the periodic propagating wave. Simultaneously this gives the sharp transitions of state that result in highly localised stripes of gene products. Pattern formation then arises from the epigenetic system as an excitable medium, with waves and periodicities as a basic pattern generator (Goodwin, 1976). Such a mechanism can act locally and propagate globally; thus there is no need to have a global distribution of variables associated with the pattern generator (e.g., diffusion gradients of gene products, which cannot occur in cellularised embryos such as the short-germ species). However, within a *Drosophila*-type syncytial embryo gene products such as those produced by *bicoid* can be distributed in a graded form and so contribute to the segmentation process. Hunding and Engelhardt suggest that this is a specialisation that is not part of the original segment generator. Their analysis indicates that the basic dynamics of segmentation involves gene products which have properties similar to those already identified in connection with the switching process for sex determination in *Drosophila,* in which transcription factors form complex heterodimers. Families of such proteins serve a variety of control functions in development. They are general-purpose regulators that give to epigenetic processes the high level of nonlinearity involved in the spatial and temporal dynamics of excitable media.

Genetic Networks

Genes lie at the interface between development and evolution. In development gene interactions reveal properties of self-organisation and relative insensitivity to developmental and mutational noise, resulting in species-specific patterns of gene expression with shared features across taxa. For evolution, gene interactions need to have properties of plasticity and change, capable of generating a large variety of dynamic sequences which can be selected and stabilised to give the diversity of epigenetic trajectories over different species. This leads to the proposition that genetic networks are complex dynamic systems capable of a type of learning. The analogies between evolution and learning have been recognised for many years, and the description of gene activity in terms of the formalisms of neural networks has been extensively developed by Kauffman (1969, 1992). However, only in the past few years has sufficient detail accumulated about gene interactions in developing organisms to begin testing genetic network models against data. Among those who have contributed to this, with particular focus on the *Drosophila* data, are Reinitz and Levine (1990), Baumgartner and Noll (1991), Ingham and Martinez-Arias (1992) and Burstein (1995). The studies of Reinitz and Sharp (1995), Reinitz, Mjolsnes and Sharp (1995) and Mjolsnes et al. (1995) introduce sophisticated and effective ways of describing genetic networks as interacting circuits with the characteristics of connectionist models (Hopfield, 1982) as used in neural network analysis.

In order to learn, genetic networks need a teacher. In biology, this role is usually assigned to natural selection, the reward for adaptive learning being survival, whereas failure results in death. However, there is nothing in such a survival constraint that explains why the developmental patterns "learned" by the genetic networks of different species should have the uniformity that has been revealed in the comparative study of epigenetic gene expression across phyla that has given rise to the concept of a 'zootype' (Slack, Holland and Graham 1993). This describes a stage in the development of all animals, from cnidarians to higher metazoa, at which a similar spatial pattern of genes of the Hox cluster (or the homologous HOM-C cluster in mammals) is expressed. Furthermore, this expression pattern correlates with a morphological stage called the phylotypic stage of development, at which the different members of a phylum converge on a common form. In the insects, this is the extended germ-band stage. All the major features of the insect body plan are then present, with the body parts in their final positions as differentiated cell groups.

Why should there be such a convergence of phylotypic form in different species, with an associated conserved pattern of gene expression? As stated by Maynard Smith and Szathmary (1995): "Although this seems to be true, the reasons are not obvious, at least to us." They then suggest, as do Slack et al. (1993), that the zootype is gratuitous, a result of common ancestry rather than convergent adaptation or developmental constraint. The implication is that the genetic network fell into an arbitrary pattern that worked, and so persisted. This view requires that all the major phyla with their convergent phylotypic stages be monophyletic. However there is no uniform consensus about this issue (cf. Minelli, 1993), and there is evidence for polyphyletic origins of some major groups just as there is for the case of pentadactyly discussed in the previous chapter. A rather different view of the 'zootype' and its origins will be considered in Chapter 9. There the proposition will be considered that the genetic network always acts within a constrained context which is a source of the order that is revealed in gene patterns. That context is the morphogenetic field with its limited set of possible sequential trajectories. A specific model of the relationships between the genetic network as a learning system and morphogenesis will then be described. For the moment I return to questions of topological constraint on biological pattern.

Topological Equivalence

Bateson recognized that in some way spatial periodicities and mirror symmetry are connected. It is instructive to examine these relationships as they have emerged in the context of insect segmentation and relate them to observations Bateson himself made in connection with limb teratologies in both invertebrates and vertebrates. In the next chapter we shall see that the same principles hold for unicellular structures. This level of universality of morphogenetic field properties is most simply expressed by topological equivalence.

Bateson observed a striking regularity in the spatial pattern of spontaneous instances of extra or supernumerary limbs, both in insects and in tetrapods. When they arise, supernumeraries usually occur in pairs and have a plane of mirror symmetry between them, a regularity appropriately described as Bateson's Rule. A morphogenetic field explanation of the phenomenon was provided by French, Bryant and Bryant (1976), who developed a model to describe a range of experimental observations arising from regeneration studies on amphibian and cockroach limbs, and *Drosophila* imaginal discs. It is known as the polar coordinate or clockface model because field values on the surface of the limb are specified in terms of a two-dimensional coordinate system of the type shown in Figure 7.18, numbers on the clock face defining circumferential values and letters defining radial values, resulting in a polar coordinate description of points on the surface. These coordinates relate in some way to morphogens defining the two dimensions of order over the limb field surface. The use of polar coordinates is not actually necessary, Cartesian coordinates corresponding to dorso–ventral and antero–posterior axes of a limb field being equally possible (Goodwin, 1976), though polars give a more convenient and elegant description for the purposes of this analysis. The proximal part of the limb (near the shoulder) corresponds to the outer circle in Figure 7.18 while the centre defines the distal tip of the limb. Imaginal discs are directly described by this geometry. Regenerative responses of the

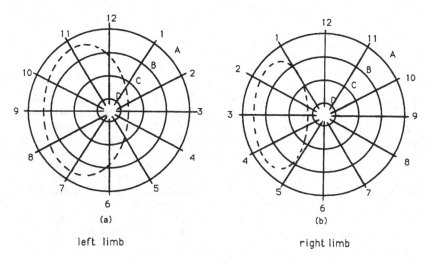

Figure 7.18. The polar coordinate model for limb pattern formation. See text for details.

limb field to experimental disturbances (cutting, grafting) are described by two rules of field behaviour: (1) a smoothing rule whereby discontinuities of state are removed wherever possible by intercalation of missing values to generate continuity of field values; and (2) a rule of distal transformation of limbs: Wherever there is a complete circle of numbers generated within the field, a limb will regenerate. The field shown in Figure 7.18a is by convention a left limb, while a right limb has the numbers running in the opposite direction (Figure 7.18b).

The model explains an impressive range of experimental results and made a number of interesting predictions, some of which were validated while others were contradicted (cf. Maden, 1982b). Its interest for us arises from an observation made by Glass (1977), who recognized that the basic properties of the model can be described in terms of its topological properties. Any closed curve drawn on a field of the type shown in Figure 7.18 can be assigned what is called a winding number or an *index*. This is defined as a property of the sequence of numbers encountered as the curve is traversed once in a standard direction (say counterclockwise) encountering a sequence of numbered radii. If these numbers define a complete circle from 1 to 12 in ascending sequence, the curve is assigned an index of 1; if the numbers define a complete circle in the opposite direction (descending sequence) the curve has index -1; and if the numbers do not complete a circle the index is 0. All the concentric circles in Figure 7.18a then have index -1, while those

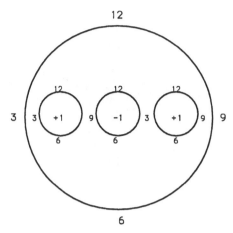

Figure 7.19. The result of the index theorem applied to supernumerary limb formation. A left limb blastema (small central circle, index −1) is grafted onto a right limb stump (large outer circle, index +1). In order for the index of the inner domain to be the same as that of the boundary, two supernumerary right limbs must be generated (the two small circles with index +1).

in Figure 7.18b have index +1. The dotted curve in Figure 7.18a also has index −1; while that in Figure 7.18b has index 0.

The rules of the model requiring that regeneration results in a globally smooth field can now be described topologically: the regenerative process must result in preservation of the initial index of the field, defined by its boundary curve. Then if supernumerary limbs arise spontaneously within an initial limb field, the sum of their indices must be equal to zero in order to preserve the initial index of the field. Therefore supernumeraries must arise in pairs of opposite handedness (+1 and −1), so there will always be a plane of mirror symmetry between them. This is the topological statement of Bateson's Rule. Note that any structure with mirror symmetry must have an index of 0, as in the case of the two mirror-symmetrical limbs.

The most celebrated result of the clockface model can now also be easily stated. Suppose a left limb blastema is grafted to a right limb stump. The right limb field has an index of +1, defined by its boundary. This must be preserved within the field. But the grafted left blastema brings into the field a circle of index −1. The only way in which the regenerating limb with this contralateral blastema can preserve a net index of +1 is to generate *two* extra supernumeraries of index +1, one to cancel the −1 blastema, giving 0, and one to give the correct index to the field. So two right-handed supernumeraries should be produced; and this is the experimental result. The regenerate ends up with three limbs, as shown in Figure 7.19, two right and one left.

The same topological analysis can now be applied to *Drosophila* segments. A colour wheel as in Figure 7.9b (*eve* vs. *hairy*) can be assigned an index since the spokes of the colour wheel carry angular coordinates from 0 to 2π, measured from a reference line through the centre. Using the same convention as in the polar coordinate model, the index of the colour wheel is $+1$ since the curve traverses a full cycle of phase angles as it is traversed counter-clockwise, which defines the antero–posterior direction along the body axis. Any closed curve in the plane can then be assigned an index according the the number of times it circles the origin. The ellipse in Figure 7.9b that circles the origin therefore has index 1, whereas the ellipse that does not circle the origin has index 0. This defines a structure with mirror symmetry, since the same sequence of numbers must be encountered in opposite directions. So we have uncovered a topological equivalence between mirror symmetries in segments and in supernumerary limbs.

What is the relation between this equivalence and homology? In the pre-ceding chapter homology was defined as equivalence under transformation involving the same generative processes. This leaves open the choice of gen-erative level at which to seek equivalence. There are aspects of segmental and limb field behaviour that involve the same generative principles, because the fields express the same properties of continuity and smoothness described by the spatial dynamics of morphogen or gene-product distributions and the mappings from these field variables to morphology. The topological equiv-alence revealed in relation to mirror symmetries arises from these high-level properties of morphogenetic fields. This defines a perfectly legitimate description of homology, and in fact the construction of a hierarchical tax-onomy of organismic forms based on generative principles makes use of these different levels of homological equivalence, as we have seen. However, it is important to be clear about the nature of the generative processes going on at different levels. Limbs do not transform into segments in insects, nor seg-ments into limbs, so their equivalence is not the same as that between struc-tures such as eyes, antennae, limbs, wings and halteres, which undergo homeotic transformation. Homeosis is one of the most striking examples of transformational equivalence, revealing the unexpected near-identity of gen-erative process underlying structures as diverse as legs, wings and eyes. These all belong within the same level of the ontogenetic hierarchy, arising from imaginal discs that are generated within segments.

However, it is perfectly possible for there to be a similarity of generative process between different ontogenetic levels. For example, insect limbs are segmented. This secondary segmentation shares generative principles with primary segmentation along the body axis. Mittenthal (personal communi-cation) has observed that the basal segment of the cockroach limb has a plane of mirror reflection which he suggests may arise from the same process as that used above to explain mutant mirror symmetries along the antero–posterior axis. So the same principles of generative order appear to be operating along the proximo–distal axis of the limb and the antero–posterior axis of the body, as expected from the similarity of structure. Segments at these different ontogenetic levels are therefore equivalent with respect to these properties, though they differ in other respects. This type of recursive operation of the same generative processes within different embryonic contexts as seen in segmentation, resulting in structures that have similarities to, but also differences from, higher level structures, is a major principle of embryogenesis. Systematic descriptions can thus be carried out of the equivalences that relate structures within single organisms as well as those between different species in terms of the same principles of comparison, as Bateson suggested. The result is that ontogenesis and phylogenesis are completely integrated, the former providing the generative principles for understanding the logical structure, the taxonomic order, of the latter. Topological equivalence of the type revealed by the application of the same morphogenetic field analysis to vertebrate and invertebrate limbs, and to unicellular structures as we shall see in the next chapter, unites the diversity of biological forms across vast taxonomic distances in terms of basic ontogenetic principles.

8

The Unitary Morphogenetic Field

Organisms are integrated entities, not collections of discrete objects.

Gould and Lewontin, 1980

Organisms as Fields

If organisms are fundamental biological entities, as argued throughout this book, then there must be a systematic way of describing them as dynamically stable wholes that undergo particular types of transformation. The logical position we adopt was succinctly expressed by the eighteenth-century philosopher Immanuel Kant in his discussion of the distinction between a machine and an organism, which deserves reiteration because of the confusion in contemporary biology over these concepts. A machine, said Kant, is a functional unity in which the parts exist for one another in the performance of a particular activity. An organism, on the other hand, is both a functional and a structural unity in which the parts exist for *and by means of* one another. The distinguishable parts of an organism (leaves, flowers, limbs, eyes, etc.) do not preexist before being assembled into a functional whole, as do the parts of a machine (see Chapter 4). Rather they emerge from the interaction of spontaneously generated differences that give rise to parts within a primary unity. This unity persists throughout the generative process and into the form that we recognize as a mature organism of a particular species. In fact it extends beyond the mature form into the next generation via the gametes, parts of the organism with the capacity to produce new wholes, since the organism of which we speak as the fundamental entity in biology is a life cycle. This is a dynamic form that is stable in the repetition of a well-defined sequence of transformations by virtue of the transmission of initial and boundary conditions and parameter values via genes and cytoplasmic organisation, resulting in the production and reproduction of the individual members of a species.

We saw in the last chapter how genes may be construed to act within this type of generative process, stabilizing sets of trajectories in a quantized dynamic system that leads to specific patterns of cell differentiation. This action of genes takes place within and contributes to a context of spatio-temporal order, the morphogenetic field, which needs to be understood as a dynamic process with distinctive properties that result in the emergence of progressively more complex spatial patterns as development proceeds. The hierachical organisation of development, described in the previous chapters, is a natural expression of the dynamics of morphogenetic fields. The objective of this chapter is to present a specific model of morphogenesis that demonstrates the nature of this process. Whereas the last chapter was focussed on the topological properties of the mapping from genes to pattern, this one concentrates on the dynamic nature of the global context within which genes act, identifying the generic properties of morphogenesis that underlie the robust qualities of development and organismic life cycles. The emphasis is on the way in which long-range order in developing organisms spontaneously generates spatially organised structural complexity. Putting the contents of the two chapters together then provides the framework for understanding both the intrinsic order and the potential diversity of organismic life cycles as dynamically transforming wholes.

The organism that will be used to illustrate these properties is the giant unicellular green alga, *Acetabularia acetabulum.* It belongs to the order *Dasycladales,* comprehensively described in Berger and Kaever (1992). This provides the opportunity of demonstrating that complex morphogenetic potential is a property of single eukaryotic cells, in accordance with classical developmental conclusions (see, e.g., Berrill, 1971). At the same time it can be used to show that morphogenetic principles are the same in the plant as in the animal kingdoms, despite significant differences of detail. So we achieve a unification across the major kingdoms and between unicellular and multicellular species.

The Life Cycle of Acetabularia acetabulum

The basic characteristics of the life cycle of *Acetabularia acetabulum* (formerly *A. mediterranea*), whose habitat is the shallow waters around the shores of the Mediterranean, are shown in Figure 8.1. Isogametes fuse to produce a roughly spherical zygote about 50 mm in diameter which breaks symmetry, producing a growing stalk and a branching rhizoid that anchors the alga to the substratum and houses the nucleus in one of its branches.

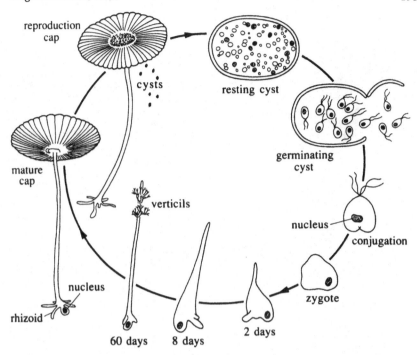

Figure 8.1. The life cycle of the giant unicellular green alga, *Acetabularia acetabulum*.

Figure 8.2. The growing and developing tip of *Acetabularia*, showing three successive verticils or whorls of laterals and a cap primordium.

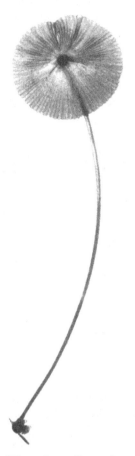

Figure 8.3. The adult form of *Acetabularia acetabulum*, a giant differentiated cell with a rhizoid containing the nucleus at the basal end of a stalk 3–5 cm long and a cap sculptured with rays at the apex.

When the stalk reaches a length of about 1 to 1.5 cm after several weeks of growth, a ring of small bumps arises around the tip, growing into a verticil, a whorl of leaflike bracts (generally referred to as laterals) that branch successively as they grow. From the centre of such a whorl the tip continues to grow and whorls are produced at irregular intervals of several days. Then, after a number of whorls have been generated, a cap primordium develops at the tip (Figure 8.2) and grows into the beautifully sculptured structure that gives the species its common name, the mermaid's cap. The laterals drop off and we have the adult form of the alga (Figure 8.3), a giant cell with a stalk about 0.5 mm in diameter and 3–5 cm in length, a cap diameter of 0.5–1 cm, and a nucleus that remains in the rhizoid throughout the developmental process. Cytoplasmic streaming carries nuclear products throughout the cell.

After about three months the alga goes into reproductive mode. The nucleus divides many times, producing thousands of nuclei that are carried up into the cap by the streaming cytoplasm. There they differentiate into haploid gametes, cysts are produced that enclose hundreds of these, and then the cell wall dissolves away and the cysts are released into the sea water. Little "trap-doors" open in the cysts, the flagellated gametes swim out and then combine in pairs to start new individuals from zygotes (Figure 8.1).

Here is a fairly typical life cycle involving a distinct process of morphogenesis leading to a particular adult morphology. The cap functions both as a photosynthetic organ and a gametophore, where the gametes are produced. But what is the function of the whorls? Why are they produced when they serve no function in the adult? They are photosynthetically active during growth, but the algae can grow perfectly well without producing whorls, or even a cap, though they do not reproduce unless caps are made. How are we to explain the production of whorls by these organisms? The Darwinist perspective appeals to history and function, as in the following explanatory scenario. Whorls were useful in an ancestor of *Acetabularia acetabulum* and have persisted in this species because of hereditary inertia and some residual utility that is sufficient to compensate for the cost of their production (at least 50% of the total resource of an individual goes into whorl production). To find evidence that supports this interpretation we examine other members of the *Dasycladales*, the taxonomic group to which *Acetabularia acetabulum* belongs. Here we do find species in which no caps are produced, the laterals functioning as the gametophores in which gametes and cysts differentiate. In fact the majority of species in this group, whose fossil members go back at least 570 million years, have whorled laterals and no caps. This is useful information in seeking an explanation of the phenemenon, but it does not itself provide an explanation for whorls in *Acetabularia*. What we need is an understanding of 'hereditary inertia', which identifies a problem rather than providing a solution. This requires going beyond historical/hereditary/functional descriptions to a study of the generative process itself.

Morphogenetic Modelling

Instead of studying algae that are going through their normal life cycle, it is convenient to work with regenerating cells. The cap can be cut off at any position along the stalk and a process of regeneration from the nucleated stalk fragment then occurs that follows exactly the same sequence of events as normal morphogenesis: after healing has occurred at the cut, a tip forms and, after a few millimetres of growth, the tip flattens and a ring of hair primordia is produced which grows into a whorl. The sequence is shown

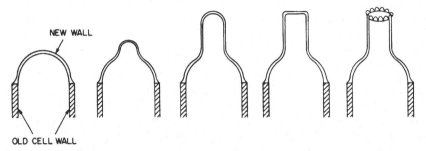

Figure 8.4. The sequence of morphogenetic changes occurring at a growing tip that gives rise to a whorl.

diagrammatically in Figure 8.4, where only the shape of the wall is shown. The cell membrane (plasmalemma) lies against the inner surface of the wall, with the cytoplasm forming a layer between the plasmalemma and the tonoplast, another membrane which separates the cytoplasm from the vacuole. The latter makes up the bulk of the cell volume, exerting an osmotic pressure on the cell wall that maintains the turgor of the cell. Tip growth resumes from the centre of the whorl and the alga produces a series of whorls before the cap is produced, as in normal growth. The cap grows, the whorls fall off, and the regenerated form is indistinguishable from the original adult alga.

Many workers have studied various aspects of the biochemistry, physiology and development of *Acetabularia* since Hämmerling discovered in the 1930s that it was a single cell which could be maintained in laboratory culture and which lent itself to detailed study because of its large size, its distinctive morphogenesis and regenerative powers. With other colleagues, I have been involved primarily in a study of the factors that influence morphogenesis and the process responsible for generating the sequence of shape changes shown in Figure 8.4, involving the formation of a whorl.

Attention was drawn to the importance of calcium in morphogenesis by studies which showed that changing the concentration of this ion in the sea water in which the algae develop causes dramatic changes of morphology (Goodwin, Skelton and Kirk-Bell 1983). These changes could be duplicated by adding to the sea water ions (Co^{2+} or La^{3+}) which block calcium channels, so the effects of reduced calcium are not simply on the mechanical properties of the cell wall but extend inside the cell. The wavelength of the whorl pattern can also be systematically altered by changing the calcium concentration in the medium (Harrison and Hillier, 1985; Goodwin, Brière and O'Shea, 1987). Calcium is known to have significant effects on the mechanical state of the cytoskeleton (Kamiya, 1981; Menzel and Elsner-Menzel, 1989). It changes the

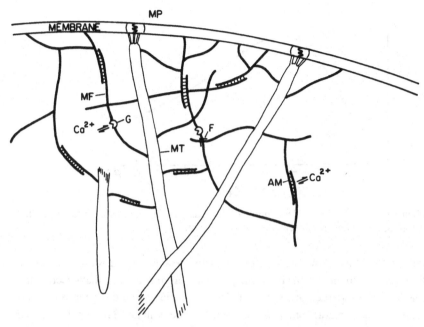

Figure 8.5. Some of the major components of the cytoskeleton and the ways in which calcium interacts with them to influence the mechanical state of the cytoplasm.

viscosity and elastic modulus of the cytopolasm by influencing the state of polymerisation of actin and tubulin, and by activating actomyosin contraction and enzymes such as gelsolin which cut actin filaments. Some of these influences are shown schematically in Figure 8.5. Comprehensive accounts of the structure and properties of algal cytoskeletons can be found in Menzel (1992).

Morphogenesis in any species depends upon a dynamic of spontaneous symmetry breaking within the developing organism whereby an initially simple structure such as the spherical zygote transforms into an adult of characteristic form. In a unicellular organism such as *Acetabularia* the most likely locus of this dynamic process is the cytoplasm. It was established by Turing (1952) that coupled biochemical reactions combined with diffusion can produce spatially nonuniform patterns of reactants which he called morphogens. These reaction–diffusion systems have been extensively used by Meinhardt (1982) and by Murray (1989) to model morphogenetic processes in a variety of organisms, and they have been considered by Harrison and Hillier (1985) and by Goodwin, Murray and Baldwin (1985) in the context of *Acetabularia* development. However, the morphogenetic effects of calcium and its influence on the cytoskeleton suggested that this system itself might play the role of primary pattern

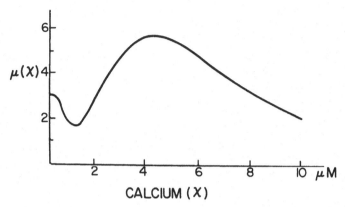

Figure 8.6. A qualitative description of the changes in the elastic modulus of the cytoplasm as a function of calcium concentration.

generator, rather than a conventional Turing reaction–diffusion process. The first step in explaining this possibility was to derive equations that describe the mechanical properties of the cytoskeleton, the cytoplasmic regulation of calcium and their interaction. This was done by Goodwin and Trainor (1985), and the coupled equations were shown to have the property of spontaneous bifurcation for particular ranges of the parameters. Within this range spatial patterns of cytosolic free calcium and mechanical strain in the cytoplasm develop from random perturbation of the system from a spatially uniform initial condition. The reason for this behaviour lies in certain basic properties of the calcium–cytoskeleton interaction, which will now be described.

Cytosolic free calcium is regulated in eukaryotic cells at concentrations of 100 nM or so by plasmalemma pumps, by sequestration mechanisms involving the endoplasmic reticulum, vesicles or vacuoles, and by binding to cytoplasmic proteins and chelating agents such as calcitonin and calmodulin. Studies of actin gels have shown that as calcium rises above 100 nM it induces gel breakdown and solation by activation of enzymes such as gelsolin. At higher concentrations calcium initiates contraction of actomyosin filaments so that the cytoplasm becomes more resistant to deformation (Nossal, 1988). At calcium concentrations above about 5 mM, depolymerization of filaments and microtubules, and the increasing activity of gelsolin, have the consequence that the cytoplasm becomes progressively solated. A qualitative description of this behaviour in terms of changes in the elastic modulus of the cytoplasm as a function of calcium is shown in Figure 8.6. This describes how calcium affects the mechanical state of the cytoplasm. There is also a reciprocal action of the mechanical state of the cytoskeleton on free calcium concentration. It is assumed that strain or deformation of the cytoplasm results in release of calcium from the bound or sequestered state to free ions. Therefore regions that happen to have elevated strain will also have elevated free calcium levels. But increased free calcium causes gel breakdown and solation. This results locally in more strain (deformation) since the cytoplasm is assumed to be under tension, producing further calcium release. The result is a positive feedback loop in the regions of calcium concentration where the slope of the elastic modulus curve as a function of calcium is negative (see Figure 8.6): a local, random increase of calcium above the steady-state level initiates a runaway calcium release and increase of cytoplasmic strain. However, this is stabilized by the effects of diffusion, which tends to reduce the calcium gradients, and also by the opposing effects of calcium on actomyosin contraction, which increases the elastic modulus and so decreases the strain (region of positive slope, Figure 8.6).

The argument also works in reverse: where calcium levels are decreased the strain is also reduced since the cytoplasm is more gel-like (higher elastic modulus) and so free calcium will be bound or sequestered, decreasing it still further. In terms of reaction–diffusion dynamics, calcium plays the role of short-range activator whereas mechanical strain is like the long-range inhibitor. The result of the interactions is that spatial patterns of calcium concentration and strain can arise spontaneously from initially uniform conditions when the equation parameters are in the bifurcation range; that is, this morphogenetic field model has the properties of an excitable medium (cf. Goodwin, 1976; Winfree, 1980). The model is qualitatively similar to that discussed by Oster and Odell (1983).

Environmental Factors

Another equally important set of influences is involved in determining the pattern of morphogenesis in addition to those acting from within the cell. These are the environmental conditions. For instance, changes in the calcium concentration of the sea water in which the algae grow can determine whether or not caps, whorls and tips can be formed, their shape and the wavelength of the whorl pattern (see Goodwin et al., 1983, 1987; Harrison and Hillier, 1985). And there is a more active aspect of environmental participation in growth and morphogenesis: electrical currents, due primarily to a flow of chloride ions into the stalk and out of the base, are a necessary accompaniment of growth (Bowles and Allen, 1986; O'Shea, Goodwin and Ridge 1990). If these currents are prevented by a current clamp, no growth or morphogenesis occurs (Goodwin and Pateromichelakis, 1979). So the dynamics of morphogenesis extends beyond the cell boundaries and includes ionic flux around as well as through the developing organism. Jaffe (1981) has argued that this is a widespread accompaniment of growth and morphogenesis. When the equations derived by Goodwin and Trainor (1985) were extended to include such flux terms, they gave properties similar to the original ones (Brière and Goodwin, 1990).

Cell Wall Dynamics

The properties of the cytoplasm described give it the characteristics of an excitable medium which can spontaneously generate spatial patterns, both stationary and dynamic – that is, propagating waves. This is sufficient to initiate pattern; but morphogenesis involves also changes of geometry. In the case of *Acetabularia* and many other developing organisms, morphogenesis is linked to growth, so the cell wall must undergo localised changes of shape together with elongation. The wall is described in the model as a purely elastic shell (about 2 μm thick) whose state changes as a function of strain in the underlying cytoplasm, which is a thin shell about 10 μm thick closely apposed to the wall, the plasmalemma separating them. This functional coupling is assumed to arise via strain-activated pumps in the plasmalemma that cause the wall to soften by excretion of protons or hydrolases. The large vacuole in the centre of the cell is an osmotic organelle, separated from the cytoplasmic shell by another membrane, the tonoplast. The vacuole exerts a pressure that is resisted by the wall. Patterns of strain in the cytoplasm are thus reflected in the elastic modulus of the wall, which undergoes elastic deformations as a result of the outward-directed osmotic pressure. A growth process was introduced into the model whereby new wall material was added wherever wall strain exceeded a threshold value so that elastic deformations led to plastic changes, in accordance with experimental evidence (Cleland, 1971; Green, Erickson and Buggy, 1971). Growth of the cytoplasm was coupled to wall growth, while vacuolar pressure remained constant. The details of the growth algorithm are described in Brière and Goodwin (1988).

Simulations of Acetabularia *Morphogenesis*

The calcium-cytogel and cell wall equations were used to simulate growth and morphogenesis of the alga. Parameters were adjusted so that the calcium-cytoskeleton equations were in the bifurcation range, making spatial patterns possible. The sequence of shape changes that we attempted to understand is that shown previously in Figure 8.4: a hemisphere, the regenerating apex,

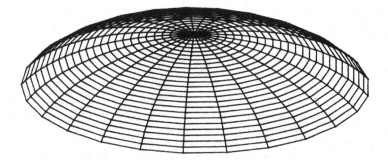

Figure 8.7. A finite-element model of the regenerating tip of an *Acetabularia* cell. The shell is constructed out of discrete elements, each of which obeys the equations of the model describing the dynamics of wall growth. A similar shell describes the cytoplasm, lying within the wall and obeying the equations of calcium–cytoskeletal dynamics.

forms first a tip which grows, then flattens prior to generating a whorl. In the model, parameter values were selected so that the characteristic wavelength of the system was small enough to allow localised deformations and pattern to arise, rather than uniform growth over the whole hemisphere. The computer simulation used a finite-element simulation of the regenerating tip, as shown in Figure 8.7. The cell wall was described as a shell of finite elements that obey the equations of elastic deformation and growth of the model, while the cytoplasm was represented as another shell of finite elements that obeyed the calcium–cytoskeletal dynamic equations. This cytoplasmic shell always took the same shape as that of the wall, against which it is apposed. Once parameters were specified to give localised deformations, they were not subsequently altered during growth and morphogenesis.

Simulations started with uniform initial conditions, on the dome. The first stage of pattern formation that typically occurred was the formation of a gradient of cytosolic free calcium that increased to a maximum at the apex. This is shown in Figure 8.8. On the left is the outline of the shape of the tip along one of the longitudinal elements of the dome, from base to apex, while the other graphs show different variables of the cytogel as a function of distance from the base. Strain is measured both in the latitudinal (solid line) and longitudinal (dotted line) elements, showing the anistropy; similarly for the elastic modulus of wall elements (incorrectly labelled elasticity). All variables start off spatially uniform (flat) and spontaneously develop a pattern, calcium rising to a maximum at the tip, as does the gel strain. The result is that the wall softens in this region and there is an elastic deformation. A three-dimensional view of this is shown in Figure 8.9: a tip is produced.

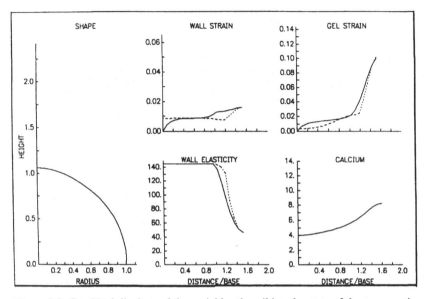

Figure 8.8. Graphical displays of the variables describing the state of the regenerating tip whose shape along a longitudinal from base (radius 1) to tip (radius 0) is shown on the left. The variables such as calcium concentration and wall elastic modulus are shown as functions of position along the longitudinal. For a tensor such as strain, with components along both longitudinal and latitudinal directions of the finite element network, the two components are shown respectively as solid and dotted lines.

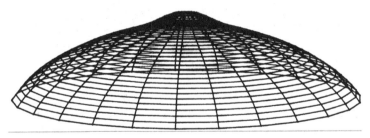

Figure 8.9. The result of the spontaneously generated increase in cytosolic free calcium concentration and strain at the apex of the regenerating dome is the formation of a growth tip, as observed experimentally (see Figure 8.4).

This is the first stage of the regenerative process, as shown in Figure 8.4. The characteristic flattening of the conical tip just prior to the appearance of the ring of hair primordia that initiates a whorl was something we had not previously understood. The model gave us an explanation. As growth occurred at the tip, with plastic changes of geometry following the elastic deformations, there was an interesting interaction between the shape generated and the dynamic behaviour of the calcium-cytogel system. After an initial stage of growth, the calcium gradient with the maximum at the tip became unstable and transformed into an annulus, the maximum level of calcium occurring away from the tip. The region of maximum cytogel strain also changed in a similar manner (Figure 8.10). Wall softening was then greatest in this annular region, resulting also in maximal wall curvature proximal to ' the tip, with consequent flattening of the tip itself, where the elastic modulus of the wall was now larger than in the annular region. As the tip grew, the amplitude of the annulus increased. (Figure 8.11). A three-dimensional colour graphic of such an annulus is shown in Figure 8.12, along with the geometry and state of the wall and the cytogel, where the colour coding shows strain.

A calcium annulus of this type was perturbed to see if it could spontaneously generate a pattern similar to that of a whorl, and the result is shown in Figure 8.13. This shows a ring of peaks of calcium that can be interpreted as the initiator of the whorl pattern. Unfortunately the finite-element program was not sufficiently robust to allow a study of the growth of such small elements, breaking the axial symmetry of the growing tip. But the sequence of pattern changes observed, namely gradient and tip formation, elongation, annulus formation and tip flattening, and the bifurcation of an annular pattern to a ring of calcium peaks, provides a very natural dynamic explanation of a basic morphogenetic sequence. These observations suggest that we may be dealing here with a symmetry-breaking cascade that is an attractor in the moving boundary process with which we are dealing, a whorl being a natural form, generated by the bifurcating sequence starting with tip formation (gradient), then tip flattening and annulus formation and finally the bifurcation of the annulus to the whorl pattern.

As the apex elongates, the annulus itself becomes unstable and intermittently transforms back to a gradient with a maximum at the tip, an annulus then reforming and growing in amplitude. This occurs with a rather irregular frequency, suggestive of the somewhat irregular frequency of successive whorl initiations as the alga grows. So we seem to have here another dynamic aspect of the morphogenetic process that occurs spontaneously, without any necessary parametric change. These aspects of the moving boundary process are again suggestive that the basic generative dynamic is robust and fundamental to this type of growth and morphogenesis. It is what Waddington (1957) called a developmental creode, but a particularly stable one. The whole morphogenetic sequence from the spherical zygote to the fully differentiated cell is following a trajectory in which it is simply breaking its major symmetries in a bifurcation cascade.

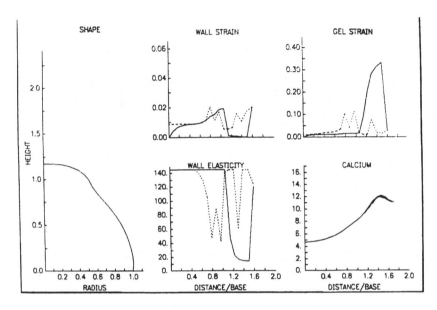

Figure 8.10. Continued growth of the apex results in a bifurcation to a calcium pattern with the maximum displaced from the tip, resulting in an annular pattern that gives rise to tip flattening.

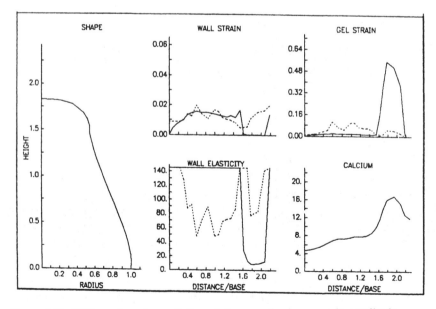

Figure 8.11. With continued growth the calcium annulus increases in amplitude.

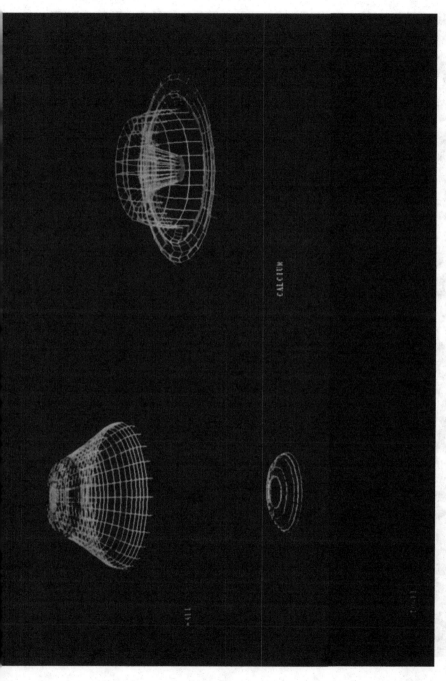

Figure 8.12. A three-dimensional colour graphic of the calcium annulus and the corresponding shape of the growing apex. This plate is available for download in colour from www.cambridge.org/9780521207430

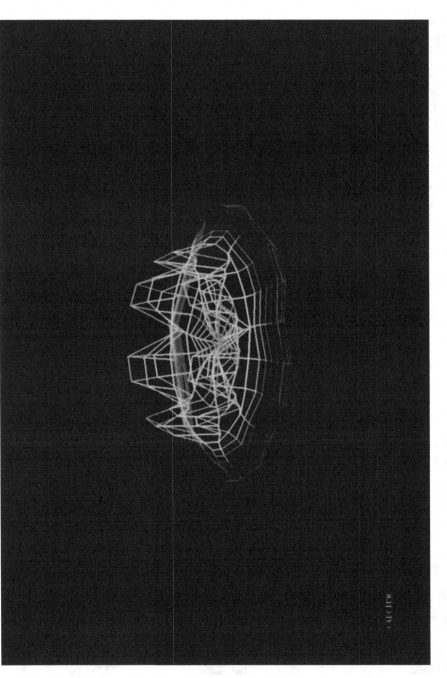

CALCIUM

Figure 8.13. The result of perturbation of the calcium annulus: a ring of calcium peaks simulates the prepattern of a whorl of laterals.

What about experimental evidence relating to the spatial patterns of calcium anticipated by the model? Using chlorotetracycline (CTC), which fluoresces in response to bound calcium, Harrison, Graham and Lakowski (1988) observed fluorescence at the growing tip of the cell. When tip flattening occurs, the fluorescence is maximal in an annulus corresponding to the region where the verticils subsequently form; and when the primordia are generated, they show the most intense fluorescence. The model predicts such results for free rather than bound calcium. However, assuming that the level of membrane-bound calcium is in equilibrium with cytosolic free calcium locally, and so reflects the latter concentration, the observations are consistent with expectations. A more direct method of measuring free calcium is to use aequorin or another calcium probe in the cytoplasm.

The model simulations have so far failed to produce a cap. This requires another pattern of growth which has been approached, but never achieved, in which lateral growth exceeds longitudinal. The conditions for cap formation appear to require more anisotropy in the strain field than is currently in the model, and it seems likely that parameter changes may be required as part of the dynamic.

The results reported here are preliminary but encouraging. What is surprising is how comparatively easy it was to find parameter values that resulted in a series of shape changes that simulate remarkably closely the morphogenetic sequences of normal growth. Tip formation and elongation can be obtained equally readily starting from a sphere rather than a hemisphere, so the boundary conditions of the latter do not impose artificial constraints on the process. However, it is necessary to make a systematic study of the range in parameter space which results in the sequence observed, in order to characterise the robustness of the model. Although it is computationally complex, it is biologically extremely simple: cell wall, cytoplasm and vacuole, modelled at their simplest. We may be looking at a basic eukaryotic morphogenerator.

The Role of the Genes

What about the role of genes in these generative processes? Their obvious influence is on parameter values. There are many of these in the model, but its behaviour is sensitive to variation in only a few of them among which the most important are the elastic modulus of the wall, the effective diffusion constant of calcium, and the restoring force of the cytoplasm. For growth and morphogenesis to occur in the model, these parameters must be in a particular range. Within this range, parametric variation results in modifications of morphogenesis: larger or smaller tip diameter, changes in the wavelength of the whorl pattern, failure of annulus and whorl formation resulting in continuous tip growth and so on. However, as noted before, there are parameter ranges in which the whole sequence of tip formation, annulus formation, bifurcation to the whorl pattern and resumption of tip growth occur. No genetic programme guides the system through its morphogenetic transitions. These are all consequences of the cycle of dynamics generating geometry and geometry-modifying dynamics, the nature of a moving boundary problem. The genes define this range, but the organising principles of the process are embodied in the spatio-temporal properties and behaviour of the cytoplasm–cell wall morphogenetic field.

These observations all refer to the behaviour of the model. In the morphogenesis of the algal cell it is to be expected that there will be differential gene influences, primarily in the form of local translation of cytoplasmic mRNAs whose protein products are involved in whorl formation. This has been demonstrated to occur in connection with cap morphogenesis, specific enzymes being produced locally at the tip during this process. These molecular activities are aspects of the morphogenetic field dynamic which may be expected to enhance and stabilise the changes of state and shape that occur as part of the natural symmetry-breaking cascade in this moving boundary process. Patterns of gene activities then reflect the capacity of the genetic network to 'learn' the constrained patterns of morphogenetic change available within a taxonomic group whose morphogenetic fields are organised according to particular principles, such as the constraints imposed by cell wall dynamics in the *Dasycladales*. This suggests that hereditary inertia in a taxon is not simply a result of gratuitous patterns established in the ancestors of a lineage, that then persist because of utility of the structures generated. Whorls do not appear to have high selective value in *Acetabularia*, so their persistence is more likely to be accounted for in terms of constraints on morphogenetic dynamics, together with the self-organising properties of genetic networks in producing stable sequences of gene expression.

A further experimental observation emphasises that the locus of spatial organizing principles is dynamics of the cytoplasm coupled to cell wall growth. *Acetabularia* cells can be made anucleate by simply cutting off the rhizoid, which contains the nucleus. If such cells then have their caps removed, they are regenerated (Goodwin, et al., 1983). The whole regenerative sequence proceeds exactly as it does when the nucleus is present. However, cells without nuclei are able to regenerate caps only once, and then only if the whole cell is put in the dark for a few days prior to having rhizoid and cap removed. This pretreatment results in the accumulation of nuclear products (messenger RNA) in the cytoplasm, which are necessary for cap regeneration. The mRNA is what generates the particular proteins required for cap formation; but it is the cytoplasm and the cell wall that are primarily involved in determining the spatial and temporal organization that defines the properties of the morphogenetic field whose activity generates the sequence of cell shapes during the regenerative process. Gene products exert their influence within the context of cytoplasmic order, the two together constituting a unity of process (see Malacinski, 1990, for an extended treatment of this theme). The evidence, both experimental and theoretical, speaks strongly against any separation of this process into stages involving an initial spatial pattern and its subsequent 'interpretation' into overt form. In relation to the model described, this simply doesn't make sense. Form unfolds spontaneously from the dynamic and feeds back upon it as a single unified process. Although in the model parameters do not need to change to get morphogenesis, in the actual organism there are very likely to be local macromolecular syntheses that change parameter values in ways that reinforce the natural dynamic. This is how gene products can act as canalising agents. Parameters are then variables in a hierarchical system, and again there is a single process, self-organising and self-generating. This argument can be extended to the whole life cycle, which of course includes relevant environmental variables. Once the dynamics of the system is clearly and sufficiently described, its process is understood at the level of the whole organism–environment cycle and its integrated properties of dynamic transformation are revealed. No distinct part is in control; the whole constitutes a self-organising system.

Hierarchical Morphogenesis

The development of form in *Acetabularia* has obvious hierarchical features. Referring to Figure 8.4, the tip of a growing or regenerating cell has a diameter of about 100 μm. The hair primordia that constitute a whorl, a transformation by secondary symmetry breaking within the spatial order of a tip, have a diameter of about 20 μm. These grow and branch via a tertiary broken symmetry into finer structures with diameters in the range of 5–10 μm. Each of these growing apices, namely, tip, primary hair, branched hair, have the same basic morphology but on a decreasing spatial scale, showing both a hierarchical relational order and a fractal-like generative algorithm. The symmetry-breaking sequence is repetitive, complexity arising by repeated dynamic bifurcations on decreasing wavelengths, involving the same type of process as that which occurs in *Drosophila* segmentation. This leads to the conjecture, which will be discussed in more detail in the next chapter in the context of a discussion of evolution, that morphogenesis is an intrinsically robust process. This means that the process is stable to many variations in the genes and in the environment so that a diversity of genotypes and of environments will result in similar morphologies. This stability is enhanced by the hierarchical structure of morphogenesis, as described in the previous chapter, and by its modularity (fields within fields, such a the limb field within the body field, and the relative independence of subfields from one another). Genes, therefore, don't have to be fine-tuned; they need only fall within certain domains of parameter space to be compatible with a particular morphogenetic trajectory. As described for *Acetabularia,* once the parameters are in this domain, morphogenesis follows its own course. This is actually the only plausible way in which species morphologies could have evolved. Morphogenetic trajectories have to be intrinsically robust to withstand the constant perturbations generated from within by genetic variations that inevitably arise from fluid genomes undergoing recombination, transposition, repetition and other genome-shuffling activities; and from without by environmental variations. The basic simplicity of the generative process in this species makes its morphogenesis relatively easy to understand as intrinsically robust, described as a moving boundary process in which geometry and dynamics interact to give a natural, stable morphogenetic sequence. In *Drosophila,* it is more difficult to uncover the underlying dynamic of segmentation because of the complexity of the genetic overwriting that has occurred during evolution and the intricacies of the gene interactions that reinforce the robust, hierarchically organized segmentation dynamic. However, the extremely widespread occurrence of segmentation as a basic morphological feature of diverse taxa leads to the conclusion that it also arises from an intrinsically robust dynamic.

Generic Forms and Taxonomy

As mentioned at the beginning of this chapter, *Acetabularia* belongs to a group of giant unicellular marine algae called the *Dasycladales.* Some species are very similar to *Acetabularia acetabulum,* with slight variations of whorl and cap morphology. An example is *A. crenulata,* in which the rays of the cap are separated so that the overall form has spherical rather than circular symmetry, the rays emanating from the apical end of the stalk like radii from the centre of a sphere. Other species have no caps, the laterals of the whorls serving as gametophores. These and other forms can be "phenocopied" in *A. acetabulum* by environmental influences; that is, this species can, without any change of genotype, be induced to take on phenotypes similar to those of other species simply by altering the environment. For instance, it was found (Goodwin et al., 1983) that changes in the concentration of calcium in the sea water caused characteristic changes of form. If the normal 10 mM Ca^{2+} is reduced to the range 3–5 mM, the algae fail to make caps so that the cells produced a series of whorls only. The laterals are then often bulbous, phenocopying species such as *Halicoryne spicata* which makes no caps, the laterals functioning as gametophores. In the *Acetabularia* phenocopies no gametes were produced in the laterals so these cells are reproductively sterile. When the calcium is reduced to 2 mM, the algae cease making whorls and the stalk simply continues to elongate. No known species has this form. At 1 mM, a bulbous tip is produced which stops growing. By altering the elastic modulus of the cell wall in the model, this bulbous form can be simulated, suggesting that there is a direct relationship between calcium in the medium and the mechanical properties of the wall, which is to be expected because of the role of calcium as a hardening agent in cell walls.

However, no systematic study has yet been made of the range of different morphologies that can be generated by parameter variation in the model. Clearly one of the objectives of this type of exercise is to find the range of forms that can be produced and to compare them with the morphologies of extant and extinct species. This could give useful information about the sizes of the domains in parameter space that correspond to significantly different morphologies, hence their robustness.

Much of the morphological variation between different species of dasy-clads arises from differences of secondary structure in the laterals – that is, how they branch, their size and shape, where the gametes are produced, and whether or not the laterals expand at their tips to produce a continuous cortex, as in members of the genera *Bornatella* and *Cymopolia* (see Berger and Kaever, 1992). The members of the group *Acetabulariaceae* share the characteristic of shedding their laterals and producing a cap which functions as a gametophore, as we have seen. The morphological character which all members of the *Dasycladales* share, the taxonomic signature of this order, is the growth of a central stalk and the production of a series of whorls of laterals, whether these are retained in the adult or not. In species such as *Bornatella sphaerica* and *Cymopolia van bossae* this form is completely hidden in the adult as a result of the secondary differentiation of the laterals to form a continuous cortex, but the typical form is clearly evident in juveniles. I shall now refer to this morphological character of the order as a *generic form*.

The taxonomic relationships of different members of the *Dasycladales* can be described by a diagram of the type shown in Figure 8.14. The space of

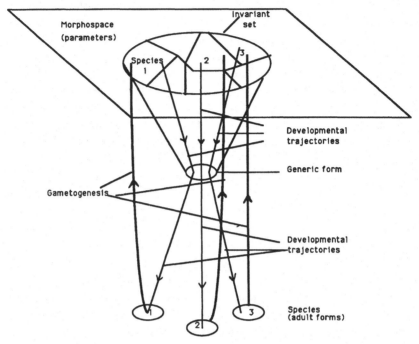

Figure 8.14. A schematic description of the relationships between genes as determinants of parameters in morphospace defining the ranges of species and the production of the generic form that is common to the group. Developmental trajectories from the species within the invariant set lead through the generic form to distinct species (adult forms) and then back via gametogenesis to the invariant set, defined by the closure of this group of life cycles.

parameters (essentially genes) that determine the particular developmental trajectories of different species is represented by the two-dimensional 'morphospace' (cf. Gould, 1991), different species occupying different regions of this space. However, all the developmental trajectories that originate within the parameter domain labelled the invariant set pass through the generic form, which is common to all members of the taxonomic group. This is the stalk with whorls of laterals, the phylotypic stage of development within this taxon. Secondary modifications of this structure result in different species with distinctive adult forms. In *Acetabularia acetabulum,* this is a structure with a cap and no whorls of laterals – the generic form has undergone considerable transformation. But the generic form is dominant throughout the development of this species, in common with all other members of the family. The adult forms of the different species then give rise to gametes through the process of gamatogenesis, which takes the life cycle back to the region of parameter space where development for that species began, and the life cycle is completed. Genes as major determinants of parameters are involved in the stability of the life cycle. If mutations take the cycle outside the region that defines a particular species, then another form is produced – either a mutant or a member of another species. And if the parameters fall outside the invariant set of the family as a whole, then what results from development (if it occurs at all) will be a different type of organism.

We are now in a position to seek an explanation of the puzzle identified at the beginning of this chapter: why does *Acetabularia acetabulum* produce transient whorls of laterals that serve no function in the adult? They do not appear to be there for their adaptive value to the species. Calling them evolutionary relicts that have persisted for nearly 600 million years of evolution because of 'hereditary inertia' just redescribes what is observed without providing an explanation. However, the morphogenetic model of *Acetabularia* development suggests that any organism whose basic organisation is that shared by all members of the *Dasycladales,* namely a single cell bounded by a thin elastic shell (a wall) within which is another shell with the dynamic properties of an excitable medium (the cytoplasm with calcium–cytoskeletal bifurcating dynamics) will naturally produce a growing stalk that periodically bifurcates to produce whorls of laterals. This form is generic for this system, where generic is being used in both its mathematical (meaning typical) and its general taxonomic sense (characteristic of a group of species). *Acetabularia* produces whorls not because natural selection has stabilised them and then cannot remove them, but because this is a natural form for this type of organism. Some species use them for reproduction and some don't, but all produce them. In this sense the form can be explained in the same way that forms are explained in physics, as consequences of the way in which natural processes are organised in terms of their generative fields. This brings biology back into a fundamental relationship with physical principles of the general type described by D'Arcy Thompson (1942) and discussed by Webster in Part I.

Phyllotaxis as a Self-Organising Growth Process

The production of whorls of laterals in dasyclads is in certain respects similar to the production of leaves in higher plants, despite the vast taxonomic distance between these groups. However, the objective of this enquiry is to look for unifying principles across the biological realm, so in this context it is tempting to examine the pattern of arrangement of leaves on a stem in terms similar to those used in studying algal morphogenesis. It has been recognised

1. Distichous (corn)

2. Whorled (maple, mint)

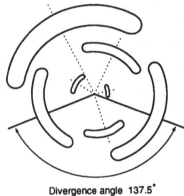

Divergence angle 137.5°

3. Spiral (ivy, lupin, potato)

Figure 8.15. The three basic patterns of phyllotaxy in higher plants.

for many years that there are only three basic types of leaf arrangement
(phyllotaxis) in higher plants (Figure 8.15). The whorled pattern of *Acetabularia* laterals, which are primitive leaves, is actually one of these though
in higher plants the number of leaves in a whorl is usually much smaller than
in *Acetabularia*. The other two patterns are distichous (leaves alternating in
position from one side of the stem to the opposite, as in maize and grasses
generally) and spiral, the majority pattern in plants. The generator of these
patterns is the meristem, the growing tip of the plant. Following on from the
previous argument, an interesting and plausible proposition is that there may
be only three basic morphogenetic trajectories for the dynamic organisation
of the meristem as a growing system with a moving boundary. This is Green's

(1987, 1989) conjecture, and he has produced extremely interesting evidence on the mechanical behaviour of the meristem to support his views. Localised growth of leaf primordia at specific positions results in a mechanical strain field over the whole of the meristem that results in particular patterns of cellulose microfibril orientation in cell walls. These orientations allow deformations to occur more easily along certain directions, accommodating cell growth and at the same time directing oriented planes of cell division. Green (1989) has presented a very plausible analysis suggesting that the three basic phyllotactic patterns are the stable morphogenetic solutions of these mechanical strain fields.

A model of phyllotaxis by Douady and Couder (1992) takes this argument a significant step forward. What they have succeeded in demonstrating is the nature of the symmetry-breaking bifurcation in the morphogenetic field of the meristem that results in the major phyllotaxis patterns, and at the same time they show how minor ones arise. They used a simple physical model to generate patterns by letting drops of a ferrofluid (a fluid with magnetic properties) fall at the centre of a disc covered with a film of oil in which the drops floated. A magnetic field polarised the drops so that they became little magnetic dipoles that repelled each other. The morphogenetic field of the meristem was thus represented by a magnetic field (see Figure 8.16). As the drops fell at the centre of the disc they experienced a repulsion from polarised drops already present, and they also were exposed to a steady magnetic field that pushed them out from the centre towards the edge of the disc. The result is that different patterns arise depending on the conditions of the experiment, but they all correspond to observed phyllotaxis patterns.

If the drops are added slowly, then by the time the next drop is added the only drop that has any influence on it is the immediately preceding one; the others are too far away to have any effect. As a result, a new drop is repelled to a position 180° away from the previous one, so that the pattern produced is like alternate or distichous phyllotaxis, as in maize or corn (Figure 8.15,1). As the rate of adding drops (equivalent to the rate of initiation of leaves in a meristem) is increased, a new drop experiences repulsive forces from more than one previous drop and the pattern changes: the initial simple symmetry of the alternate mode gets broken and a spiral pattern begins to appear. It takes a while for the system to settle on a steady pattern, the time required being dependent on the rate of adding drops. If this is rapid, so that there is strong interaction between drops, then a stable pattern emerges rapidly and successive drops quickly settle into a divergence angle of 137.5°. An example of this is in Figure 8.17, which also shows the rapid convergence of the angle

(a)

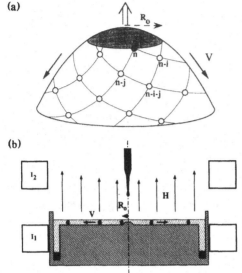

(b)

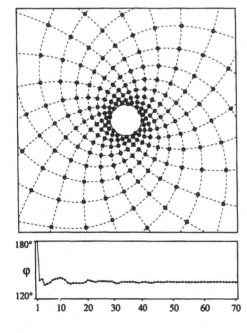

Figure 8.16. A description of the model (b) used by Douady and Couder to simulate leaf production on a meristem (a) (From Douady, S. and Couder, Y. (1992) Phyllotaxis as a physical self-organised growth process. *Phys. Rev. Lett. 68*, 2098).

Figure 8.17. A spiral pattern with the Fibonacci number pair (13,21) generated by the Douady and Couder model, with convergence of 0 on the value 137.5° starting at 180°, shown below (From Douady, S. and Couder, Y. (1992), Phyllotaxis as a physical self-organised growth process. *Phys. Rev. Lett. 68*, 2098).

between successive drops on 137.5°, starting at 180°, which is the angle that will always form between the first two drops. The actual number pair that describes the spirals depends on the rate of addition of drops, the one shown having the values (13,21) – 13 spiral arms going one way, 21 the other. Douady and Couder found other divergence angles These numbers belong to the Fibonacci series where connection with spiral phyllotaxy has long been recognized such as 99.502°, 77.955° and 151.135°, which are among the minority classes that are occasionally observed in plants. These minority patterns turn out to be significantly less accessible than that generated by a 137.5° divergence angle. The model revealed that there is a major trajectory that leads from an initial distichous pattern, through a sequence of divergence angles that initally decrease rapidly from 180° and then converge on 137.5° through a pattern of damped variations, as shown at the bottom of Figure 8.17. The trajectory produces transient spiral patterns that settle on the one that is most commonly found in nature. The other spirals are less robust than the dominant one.

This important conclusion was demonstrated by Douady and Couder in the following way. The quantity they used as the parameter controlling the transition from alternate to spiral phyllotaxis is a dimensionless number which they define as $G = V\,T/R_o$, where V is the rate at which drops are moving away from the centre of the disc under the action of a steady magnetic field, T is the period between addition of drops, and R_o is the radius of the region that corresponds to the centre of the meristem around which the leaf primordia are generated. As T is decreased, so that drops are added in more rapid succession, G decreases and the transition from alternate to spiral phyllotaxis occurs when G is *less* than a critical value which they designate as $G_{1,1}$. The (1,1) here is the first pair of numbers in the Fibonacci series (1,1,2,3,5,8,13,21, . . .) and corresponds to distichous phyllotaxis (one spiral only joins successive leaves or drops, and this spiral can be drawn in either direction, right- or left-handed). For values of G less than $G_{1,1}$, spiral phyllotaxis begins, and as G is decreased continuously the normal sequence of Fibonacci spirals is generated, corresponding to successive pairs of numbers, with fairly abrupt transitions between them. The divergence angles corresponding to successive values of G are shown in the diagram in Figure 8.18. The main curve (triangles) starting at $\phi = 180°$ converges towards $\phi = 137.5°$, with oscillations about this value as G decreases and the phyllotaxy number changes systematically. Two series are shown, corresponding to different energy functions describing the strength of the inhibition between the drops ('leaves'). The exact values of $G_{1,1}$ at the transition from distichous to spiral phyllotaxis differ, and so do the details of the curves, but their basic

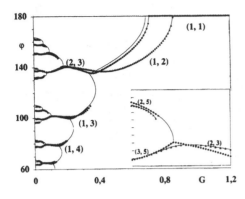

Figure 8.18. The sequence of bifurcations starting at (1,1) phyllotaxis (180° divergence angle) and proceeding through successive Fibonacci number pairs, the main sequence converging on the divergence angle of 137.5° as G decreases independently of the energy function used to describe the field (triangles or circles). Secondary sequences converging on other divergence angles are shown as discontinuities. Insert: detail of the main bifurcation sequence and a secondary sequence (From Douady, S. and Couder, Y. (1992), Phyllotaxis as a physical self-organised growth process. *Phys. Rev. Lett.* 68,

properties remain unchanged – that is, the model is robust to differences in the details of the morphogenetic field. However, there is a quantitative relation that does not change, and it is this that establishes the transition as a symmetry-breaking bifurcation (i.e., a change from one stable pattern to another). Douady and Couder show that the divergence angle varies as

$$180° - \phi = (G_{1,1} - G)^{1/2},$$

which defines a square root relationship between ϕ and G in the neighbourhood of the transition. This identifies it as a symmetry-breaking bifurcation. The implication is that the robust sequence for plant meristems is precisely the major Fibonacci series that is observed in higher plants – that is, the arrangements of leaves observed in nature are the generic forms that result from a self-organising, robust morphogenetic process. The angles other than $\phi = 137.5°$ can occur under particular circumstances, and are occasionally observed in nature, but they are minority classes that are not reached by the main sequence. So higher plants generate this aspect of their form simply by doing what comes naturally – that is, by following robust morphogenetic pathways to generic forms, as was proposed for the production of the whorled pattern of laterals in the *Dasycladales*.

There is one pattern missing from this description: whorled phyllotaxis. This would be obtained if more than one drop was added at any one time, with $G > G_{1,1}$ so that only the influence of the previous cluster of drops was experienced. Then each cluster of drops would take up positions at maximal distance from each other, and successive clusters would be arranged in the spaces of the previous cluster, resulting in a whorled pattern. So all the patterns can be generated simply by changing growth rates and numbers of leaves generated at any time, which are presumably the main parameters that differ between plant species. The neighbourhood relations of the different members of the sequence are clearly defined by their proximities in the symmetry-breaking process described in Figure 8.18 and the transitions to the possible minority Fibonacci number pairs (e.g., (1,3), (1,4)). These are all transformations of one another under change of parameters and initial conditions defining the possible set for leaf patterns in higher plants.

Over 80 percent of the 250,000 or so species of higher plant have spiral phyllotaxis. This is also the dominant form generated in the model, which identifies it as the most probable form in the morphospace of possible phyllotactic patterns. So we get an interesting conjecture: the frequency of the different phyllotactic patterns in nature may simply reflect the relative probabilities of the morphogenetic trajectories of the various forms and have little to do with natural selection. That is to say, all the phyllotactic patterns may serve well enough for light gathering by leaves and so are selectively neutral. Then it is simply the sizes of the domains in the generative space of these generic forms that determines their differential abundance. This is not to deny that the forms taken by organisms and their parts contribute to the stability of their life cycles in particular habitats, which is what is addressed by natural selection. It is simply to note that an analysis of this dynamic stability of life cycles can never be complete without an understanding of the generative dynamics that produces organisms of particular form in the first place, whose intrinsic stability properties may play a dominant role in determining their abundance and their persistence. The objective is not to separate these different aspects of life cycles, but to unify them in a dynamic analysis that puts natural selection into its proper context – in no sense a generator of biological form, but one of the factors involved in the stabilisation of form.

Morphogenetic Fields and Cortical Inheritance

From the analysis presented, we can say that organisms are dynamic forms with memory (inheritance). They also embody a particular kind of agency implying a distinctive type of causation, which will be considered at the end of this chapter. However, here the focus is on morphogenetic fields, which have the generic property of a hierarchical unfolding of complex order during the developmental phase of the life cycle. By ensuring that parameter values fall within a certain domain, genes contribute to the stability and repeatability of the life cycle of any species, which is biological memory or heredity.

Other sources of stability come from the environment and from the boundary conditions of the morphogenetic field, which influence the selection of particular epigenetic trajectories. Examples of inherited boundary conditions come from the ciliate protozoa, whose cortical inheritance is widely recognized as an aspect of the hereditary process. At one level, this depends upon the local organization of nucleating structures that guide protein assembly into the characteristic patterns of unit territories bearing cilia on the cell surface. These are normally organized into rows, called kineties, running from anterior to posterior pole. The uniform polarity of these kineties in normal cells results in a uniformity of ciliary beat pattern whereby the organisms move through water. If a row is inverted by surgery then this altered pattern is inherited, all the progeny having the corresponding row inverted (Sonneborn, 1970; Ng and Frankel, 1977). No change of genotype is involved. As discussed in Chapter 6, such phenomena of inheritance are due to the continuity of the soma in asexual reproduction so that alterations in boundary and initial conditions of the morphogenetic field are propagated, like defects in a crystal. Figure 8.19 shows this continuity of cell structure during the later phases of reproduction in *Tetrahymena,* when the oral apparatus is being duplicated and the membranelles are forming just prior to cell division. The rows of cilia constituting the kineties are shown running from pole to pole of the cell. A schematic close-up of normal unit territories, the units of cortical structure with their cilia, is shown in Figure 8.20, along with an inverted row. Unit territories are generated as the cell grows by a local assembly process, as illustrated in Figure 8.21, which explains why an inverted row is stably propagated from generation to generation. However, as is evident from Figure 8.19, there are longer-range influences than unit territory assembly: the oral apparatus of the progeny cell is assembled under the influence of the parent OA at a distance of some 20–30 μm beneath it, located adjacent to the same kineties that are confluent with the original OA. Also there are intermediate levels of spatial order, as we shall now see.

The notions of hierarchical generative order, global and local aspects of morphogenetic fields, apply to pattern formation in these unicells as well as to any other developing organism. This is revealed most clearly by mutations that affect different levels of the morphogenetic hierarchy, just as they do in *Drosophila.* For example there are mutants such as *janus* that stabilize a mirror-symmetrical form in which a second oral apparatus (OA) appears with the opposite handedness to the normal, as defined by the spatial pattern of the membranelles of the oral apparatus (Frankel and Jenkins, 1979). However, there is a conflict between the asymmetry of membranelle orientation in this second OA and the asymmetry of the unit territories with their cilia

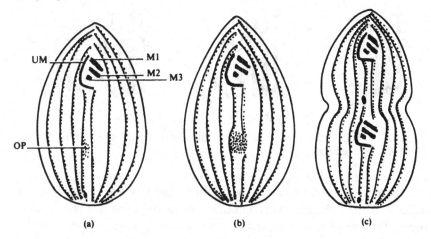

Figure 8.19. (a) The arrangement of kineties on *Tetrahymena* and the membranelles (UM, M₁, M₂, and M₃) that define the oral apparatus (OA). An oral primordium (OP) appears below the equator, adjacent to the same kineties as the original OA, and develops into a new OA prior to cell division (b and c).

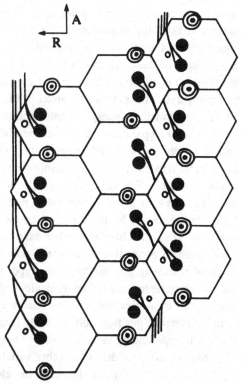

Figure 8.20. The detailed structure of the unit territories of a kinety, with one row inverted. The cilia are anchored to basal bodies which are asymmetrically located in the unit territories so that a distinct handedness can be defined (R = cell's right; A = anterior).

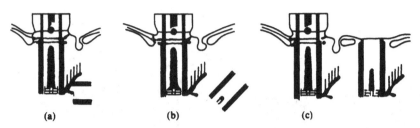

Figure 8.21. Stages in the construction of a unit territory adjacent to an existing one, which acts as a template.

that make up the membranelles, whose handedness is not altered in these mutants. The result is a rather imperfect structure. What this reveals is a typical hierarchy of morphogenetic influences, the long-range field that determines the position, number and handedness of the OAs and the local field that acts within and between unit territories. Another mutation, *disorganised,* disturbs the latter field so that the normal spatial patterning of unit territories into kineties is disrupted (Frankel, 1979). In between these is an intermediate level of field organisation, revealed by mutants in which the normal number of membranelles in an oral apparatus is altered (Frankel et al., 1984), but the number of OAs, the orientation of the membranelles in the OA, and the local organization of kineties are all normal. A detailed analysis of the basal body composition of these mutant OAs showed that the membranelles constituting them, which could be fewer or greater than the normal 3, had modified structure compared with normals. Frankel et al. (1984) conclude that the whole set of membranelles undergo transformation as a unified field, resulting in a range of oral apparatus forms with membranelles varying from 2 to 5 in number, with structure and composition intermediate between the normals. In his extremely interesting and comprehensive volume on the subject, "Pattern Formation," Frankel (1989) compares the membranelle series with Bateson's studies on serially duplicated parts and recalls his conclusion that "the whole Series of Multiple Parts is bound together in one common whole" (Bateson, 1892). So here is another example of a transformation set arising from common generative processes in which the pattern produced is not the result of a naming process by a genotype, interpreting a prepattern. Rather, the morphogenetic field is hierarchically organized over different spatial domains (unit territories and kineties, membranelle field of the oral apparatus, field influencing position and orientation of secondary oral apparatus production in reproducing cells with one or two OAs). Gene products can influence each of these fields, as can other variables. A primary candidate for this is, once again, calcium, as shown in the elegant model of Le Guyader and

Hyver (1991) demonstrating how calcium waves are likely to be involved in the duplication patterns of kineties in *Paramecium*.

Apart from its hierarchical nature, morphogenesis in the ciliate protozoa and perturbation by mutation and environmental stimuli reveal other familiar properties. The mirror-symmetrical mutant phenotype *janus* is of the same general class as those described in *Drosophila* and is subject to a similar analysis. Frankel (1989) has explicitly analyzed the parallels between mirror symmetries in ciliates, in *Drosophila*, and in limbs, covering precisely the same ground as I have, though without developing a specific model for the *Drosophila* mutant phenotypes. His primary focus is to use the insights of the polar coordinate model to understand a fascinating set of observations that he and his colleagues have made on abnormal ciliate phenotypes.

The *janus* mutant phenotype, with its two oral apparatuses in mirror-image display, is an example of a heteropolar doublet. Such a form can be phenocopied by perturbation of wild-type cells (e.g., heat-shock, surgery) and the mirror-symmetric structures are then transmitted to the progeny by cortical inheritance. Another abnormal form is the homopolar doublet, a cell with two OAs both with normal handedness, again produced by perturbation. Such cells have a greater diameter than normal and they tend to regulate back to normal size, losing one of the OAs in the process. A transient structure often appears during this regulation. The two OAs in the doublet are usually located on meridians at less than 180°, and within this narrower sector of the cell there appears a third OA of opposite handedness. These approach one another as cell size and kinety number decrease towards normal and then two fuse and annihilate, leaving a normal singlet. By a simple extension of the index theorem used in the previous chapter, this apparently paradoxical behaviour becomes clear. Frankel (1989) does it in terms of an extension of the clock-face model, as shown in Figure 8.22.

Figure 8.22a represents a homopolar doublet, with numbers 1 to 10 (instead of 1 to 12) used to describe the normal circumferential field values. Using the conventions introduced in Chapter 7, the index of the circle is -2, since the full sequence of numbers (1–10) is encountered twice in reverse order when the circle is traversed counterclockwise. (Frankel, 1989, uses clockwise as standard direction and calls the index the winding number, following Winfree, 1980. I use counterclockwise as standard because this is the usual measure of angle in polar coordinates.) A normal cell has index -1. The transition from -2 to -1 can be achieved by introducing a mirror-symmetrical structure in the region where the coordinates are crowded together, representing the reduction in kinety number as the cell regulates towards normal size. Frankel describes this in terms of a sudden reversal of

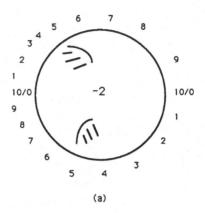

(a)

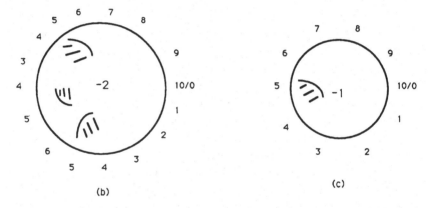

(b)   (c)

Figure 8.22. (a) A representation of a homopolar doublet of *Tetrahymena* with the circumferential field described by the numbers 1–10, as in the clockface model. Using the same conventions as in Chapter 7, the index of such a cell is −2. In regulating back to normal index −1, a third OA of index +1 can appear transiently (b) and then two OAs of complementary index fuse and annihilate, leaving a normal cell (c).

field values in the range corresponding to an OA (5 in the diagram; see Figure 8.22b). The mirror-image domain then annihilates by fusion of field values, resulting in a normal cell (Figure 8.22c). The reversal could occur anywhere around the circumference, corresponding to structures other than an OA that appear transiently, and those are also observed. But only the OA provides such clear evidence of mirror reflection, because of its structural asymmetry.

The colour wheel model applied to this phenomenon is straightforward but uses a slightly different representation. The circumferential field of a homopolar doublet is described as two turns around a colour wheel with its phase-

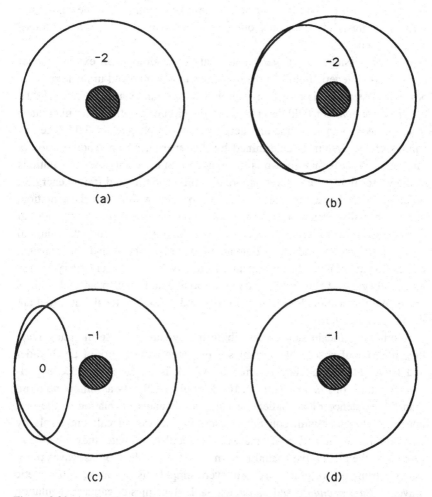

Figure 8.23. Application of the colour-wheel model to doublet regulation in ciliates. A doublet is described by two turns around the colour wheel, with index −2 (a). During regulation, reduction in the circumferential field corresponds to contraction of one of the wheels (b). When this contraction results in a crossing of the null disc (c) a mirror-imaged pattern appears (e.g., a pair of mirror-symmetrical OAs) which then disappear with further contraction, leaving a normal cell of index −1 (d).

less disc at the centre, as shown in Figure 8.23. In terms of the afore-mentioned conventions, the doubly wound wheel has index −2. Reduction in cell size then corresponds to a shrinkage of one of the wheels as shown in Figure 8.23b. When this distortion results in the curve crossing the singularity, a mirror-symmetrical domain appears (Figure 8.23c) corresponding to Figure 8.22b with the pair of OAs in mirror symmetry. The curve then shrinks progressively until only one turn of the wheel remains, representing a normal cell.

What is missing from these topological descriptions is an explicit model of the morphogenetic field and the dynamics that is involved in the regulation process. A step in this direction is provided by Brandts and Trainor (1990a, b). They use a vector field description of the circumferential order in ciliates, which allows them to define an energy of the morphogenetic field. The dynamics of the system is determined by this energy, which tends towards a minimum. What is then shown is the existence of different possible solutions of the vector field (circumferential order), corresponding to different energies. Solutions of higher energy, corresponding to fields with reversed orientation, do not normally occur; but as homopolar doublets regulate towards normal by decreasing their circumference, there are conditions in which the minimal energy state corresponds to a domain of pattern reversal and the transient mirror-imaged oral apparatus appears. In this type of treatment everything is made fully explicit and unnecessary conceptual details disappear. Field values are spatially distributed with global order and define the local states that are the observed pattern.

Another significant step in describing the dynamics of the morphogenetic field in terms of observable variables is provided by the work of Le Guyader and Hyver (1991), already referred to, which is based on detailed experimental studies of *Paramecium* (cf. Iftode et al., 1989). By assuming no more than the existence of a gradient of calcium channels in relation to the oral apparatus, together with equations describing control of calcium levels by calmodulin and the action of kinases that modulate its state, they are able to reproduce the distinctive spatial pattern of kinety duplication as waves propagating from the original OA. An even simpler model of morphogenetic waves in *Paramecium* based on the excitable dynamics of calcium regulation has been presented by Laurent and Fleury (1995).

This type of model succeeds in describing morphogenesis in a fully dynamic manner that includes wave propagation as an intrinsic aspect of spatial patterning, as expected in excitable media (cf. Goodwin, 1976, 1994). In a discussion of the implications of a mutant called *hypoangular* in the ciliate *Tetrahymena thermophila*, Frankel et al. (1993) recognise this as a possible resolution of the paradoxical properties of this mutation in relation to static two-stage models. By appealing to a wave-propagation process involving calcium as a possible morphogen they recognise a diversity of dynamic consequences that could explain the range of forms that can be generated in the mutant. They also suggest how the observed mutual dependence of morphogenetic field values along both the antero–posterior and the circumferential axes of the organism may be understood in terms of participation of calcium in both patterning axes. These proposals go a long way in the direction of revealing the unitary nature of morphogenetic fields in developing organisms, and the use of the same principles of organisation in unicellulars and in multicellulars in terms of a fully dynamic treatment of development.

Causation and Powerful Particulars

Before concluding this chapter on morphogenetic fields, it is important to recall the issues of causality that are involved in their dynamics. Organisms have been defined in this work as entities organised dynamically by developmental or morphogenetic fields. The claim has been made (see Chapter 4) that morphogenetic fields are powerful particulars; that is, they are things with a particular type of agency or causal power. The theory of causation used here comes from the work of Harré and Madden (1975), who go back to Faraday's (1857) definition of force in terms of *powers,* quoting him in their book *Causal Powers.*

What I mean by the word [force] is the *source* or *sources* of all possible actions of the particles or materials of the universe: these being often called the *powers* of nature when spoken of in relation to the different manners in which their effects are shown. (p. 175)

Harré and Madden summarise this view of causal agency as follows:

Causality is to be understood on the model of the active production of effects by powerful particulars, such as magnets, falling stones, compressed gases, stressed metals, and the like. The model for causal concepts thus shifts from the essentially inert items of the billiard ball paradigm to the essentially active items of the set of paradigms listed above.

Thus process in nature is to be understood in terms of the actions of things that are powerful particulars. What specific type of powerful particular are organisms, defined as morphogenetic fields?

Emmet (1992) has made the important point that organisms are more than simply dynamic forms or processes, because they have the quality of acting and being acted upon as a unit. She uses the term thinglike to characterize this property of integral agent, and describes organisms as "things-in-process" to designate their dynamic natures. The implication is that the causal powers of organisms have a distinctive property that results in continuous change or transformation, in accordance with the dynamic constraints that characterise organisms. The problem of biological form involves identifying and describing the characteristics of "things-in-process" as morphogenetic fields, and identifying equivalent morphogenetic processes which define homologies at the different hierarchical levels of these fields. Each of these involves the expression of a particular developmental competence, a power characteristic of a kind of field that generates a characteristic type of form: a vertebrate eye, a tetrapod limb, a digit, a segmented body, a leaf arrangement (phyllotaxis), a petal, a hair and so on. These competences result in equivalence classes of structures and these classes can be related to one another hierarchically, as the class of digits is contained in the class of tetrapod limbs, and tetrapod limbs within the class of appendages to which fish fins also belong.

The evidence presented suggests that organisms obey similar generative principles whether they are unicellular or multicellular. Organisms are dynamic wholes that undergo transformation, but they are complex entities both in composition and in the diversity of spatial and temporal patterns which are available to them. Genes can be regarded as specifying parameter values of the developmental dynamics under certain conditions; but because the system is hierarchically organised, with different time scales of change for different processes, these parameters are themselves subject to change with characteristic relaxation times and become dynamic variables. A full description of this type of hierarchical dynamic system, and the identification of the kinds of generic order that can emerge, is by no means a simple task. However, this type of investigation is now a part of the burgeoning area of interdisciplinary research that has come to be known as the sciences of complexity, the study of emergent order in complex systems, within which the work described here belongs. The next chapter explores some of the wider biological implications of this context.

9

A Generative Biology

The few who wonder if a full recognition of the role played by development will threaten the basic principles of Darwinian adaptationism can see how far one of the pioneers of genetics was willing to carry this line of argument.

P. J. Bowler, Foreword to W. Bateson's "Materials for the Study of Variations,"
Johns Hopkins University Press republication (1992)

Development and Evolution

Introduction

There are many different ways of approaching a subject as vast as biology. Darwin's *Origin of Species* cast it in an historical and functional mould, which tends now to be the predominant mode of thinking about the phenomena of evolution. However, there is a complementary view which has been stressed throughout this volume which is concerned with the origins of order in biology. These are not necessarily in conflict, as stressed by Stuart Kauffman in a recent book with this title (*The Origins of Order: Self-Organization and Selection in Evolution*, 1992); though there is an as yet unresolved tension between principles of self-organisation and natural selection (Depew and Weber, 1995). A dynamic analysis involves both pattern and process, a description of the generative (causal) origins of the forms under consideration as well as a study of contingent effects (initial and environmental conditions) on stability and change. Now that the tools of nonlinear dynamical analysis have reached the degree of sophistication necessary to explore the properties of complex systems, some extremely interesting and often counter-intuitive behaviour is emerging. Indeed it is emergent phenomena that can be said to be revealed in such systems, new levels of simple dynamic order arising from extreme complexity; and conversely, complexity arising from simple dynamic order. The latter has been made dramatically evident in the phenomena of deterministic chaos and fractals, whose explosive growth followed the pioneering studies of those such as May (1976), Feigenbaum (1978) and Mandelbrot (1977). Order from complexity is a theme that has been developing rapidly from its roots in the study of nonlinear systems operating far from thermodynamic equilibrium, resulting in patterns such as Benard convection cells, Turing reaction–diffusion patterns and the dynamic order that emerges in what are now known as excitable media (Goodwin, 1976; Winfree, 1980, 1987; Glass and Mackey, 1988). The latter are remarkable in that similar patterns of standing waves, propagating waves (concentric circles and spirals), and chaotic behaviour arise in systems as apparently different as the Beloussov-Zhabotinsky reaction and the cellular slime mould (Nicolis and Prigogine, 1987), hearts (Winfree, 1987), brains (Skarda and Freeman, 1989) and ant colonies (Miramontes, Sole and Goodwin 1993). The key to this unexpected uniformity of emergent order in complex systems appears to lie in the way they are organised rather than their composition. Their properties of spontaneous bifurcation from spatially uniform to nonuniform patterns that characterise excitable media, combined with principles of symmetry breaking that follow constraints arising from the dynamic and geometric symmetries of the system, result in symmetry-breaking cascades of the type already considered for tetrapod limbs, *Drosophila* segmentation, *Acetabularia* whorl formation and plant phyllotaxis. The general mathematical principles of such behaviour have been clearly described by Stewart and Golubitsky in their book *Fearful Symmetry* (1992), while the implications of the new perspective regarding general issues of intelligibility and order in science are explored by Cohen and Stewart in *The Collapse of Chaos* (1994). What is developing from these areas of study is an understanding of some general principles of complex systems that begin to illuminate the origins of biological order (Goodwin, 1994a).

My goal in this chapter is not to elaborate on these dynamic principles, exciting as they are, but to explore the general research programme that arises from their application to organisms as developing and transforming systems of a particular kind and its implications concerning the interpretation of evolutionary processes. What is involved here is an exploration of what Sober (1993) has described as "a theory of biological kinds" whose purview is a study of the more robust and less contingent aspects of biology, a programme also advocated by Gould (1977, 1991).

Morphogenesis as an Intrinsically Robust Process

Development involves many constituent processes occurring simultaneously at different levels – molecular synthesis, gene activations, spatial patterning of substances, cell interactions and morphogenetic movements. These diverse activities are orchestrated into the remarkably coherent transformations that generate the basic body plans of different phyla, and the unfolding of morphogenetic detail that results in species morphology. Such phenomena might conceivably arise in two very different ways. Either the constituent processes are unlikely concatenations of events that are channelled into coherence by close control over the range of parameter values that make the coordinated transformations possible; or there are dynamic principles of self-organisation at work in development such that coherent states of spatio-temporal order are highly probable, the generic states of this class of system. I shall now examine the second alternative, whose consequences concerning our understanding of evolutionary processes could be considerable.

The dynamical analysis of *Acetabularia* morphogenesis presented in the last chapter provides an example of what is meant by an intrinsically robust process, one in which the dynamic organisation of the system resulting from the interactions of the constituent processes results in a morphogenetic sequence that does not require continuous fine tuning of parameters to guide the system to a particular state. It was shown there that, once parameters were adjusted so that the intrinsic wavelengths of emergent patterns were smaller than the size of the initial regenerative domain so that detailed structure could be generated, the system had a natural tendency to pass through a sequence of shape changes that are qualitatively similar to those observed during regeneration: tip formation, growth, tip flattening, whorl formation and then a sequential repetition of this process. Localised gene activity via controlled translation of cytoplasmic mRNAs accompanies development and morphogenesis in algal cells. It seems reasonable to propose that gene activity will occur in such a way as to reinforce dominant trajectories, resulting in the highly robust morphogenetic sequences that generate whorls of laterals as generic forms throughout the *Dasycladales*. That is to say, genetic networks learn to stabilize generic forms. Differences in these structures between species are expected to correlate with differences in gene activities superimposed on common patterns.

Goodwin, Kauffman and Murray (1993) and Goodwin (1993) have presented a general case for morphogenesis as an intrinsically robust process, based upon nonlinear dynamics and the way in which an initial pattern of broken symmetry biases subsequent symmetry-breaking events, resulting in an epigenetic cascade that tends to unfold in a particular sequence. We have seen the operation of such a cascade in the hierarchical unfolding of the segmentation process in *Drosophila,* described as a sequence of harmonics of a global generative field that show up in the spatial patterns of segmentation gene products. These products amplify and stabilize the particular sequence of symmetry breakings that occur in this species. Tetrapod limb formation is another example of a robust cascade, each stage of the pattern-forming process defining the conditions for the next set of elements, in proximo–distal order. The *Acetabularia* example shows that the hierarchical character of pattern generation characterizes unicellular as well as multicellular morphogenesis, successive bifurcations occurring on finer spatial scales to generate nested, ordered complexity.

Newman and Comper (1990) have also expressed this perspective very clearly:

In contrast to genetic programming hypotheses, which can potentially account for any imaginable form or pattern, we have proposed an analysis of development that embodies generic mechanisms that act on tissues and their components, either during the modern ontogeny of an organism, or on its ontogeny sometime in its evolutionary history. Changing patterns of gene expression during development can drive morphogenesis and pattern formation by making tissues responsive to fresh generic effects. Genetic change during evolution can act to conserve and reinforce these morphogenetic tendencies, or in rare instances, set phylogeny on a new path by establishing susceptibility of the embryo or its tissue to different generic forces. Such generic–genetic interactions will not give rise to all conceivable forms and patterns that may be constructed from living cells and biological macromolecules. They may nonetheless provide a concrete account of why organisms achieve the particular variety of forms with which we are so familiar. (p. 15)

The Eye – Improbable or Generic?

It is worth considering another example now because it illustrates very clearly the shift of perspective that is being proposed. When Darwin contemplated the design of the vertebrate eye, he had very mixed feelings. On the one hand, he was filled with awe at this remarkable result of the evolutionary process; on the other, he found it an enormous challenge to his theory of evolution by natural selection – it gave him a cold shudder, he said. How could random variations ever have conspired to produce a functional eye in the first place, the initial step required before natural selection could get a grip on such an apparently improbable organ and subsequently refine it into a sophisticated visual system of the type we find throughout the vertebrates as well as those in invertebrates such as gastropods, cephalopods, crustacea and insects? What is even more extraordinary is that this type of organ has evolved independently in at least 40 different taxa (Salvini-Plawen and Mayr, 1977). Eyes seem to pop up all over the evolutionary map, and each time they present the same challenge, provoke the same Darwinian shudder: how could random independent events ever generate such inherently improbable, coherently organised processes as those required to generate a functional visual system in the first place? The very fact that similar, though by no means identical, processes have occurred many times in evolution already suggests that embryos make eyes as easily and as naturally as *Acetabularia* and its relatives make whorls, as insects make segments, and as limb buds make tetrapod limbs. To see this, it is necessary to review some basic aspects of embryogenesis. The argument will be restricted to vertebrates, but it applies by extension and variation to embryogenesis in other phyla.

Morphogenesis occurs by the iteration of a repertoire of basic cellular processes whose consequences vary as cell properties and their contexts change. As described in Chapter 8, in plants the generation of form is always accompanied by growth because plant cell walls are relatively rigid and it is only by plastic deformations resulting from new wall synthesis that shape changes occur, whether in single cells as in *Acetabularia* or in the meristem of a higher plant where differential growth and oriented cell divisions determine morphogenetic patterns. So plant form always results from outgrowth, and the full complexity of the organism's shape is visible to the external observer (including the roots, so long as the plant is grown in liquid or other medium that makes them visible). However, animal embryos can develop complexity by either an outward or an inward deformation of sheets of cells which, lacking walls, are flexible and can change shape with or without growth. This also makes it possible for animal cells to actively migrate over surfaces, either as organized sheets or as single cells, or to send out processes such as axons that grow over surfaces in an organised, directed manner. The result is that animals can develop great internal complexity as well as intricate external patterns. But, as we shall now see, this is achieved by basically the same type of cellular organization as that which operates in plants, though now with the freedom of movement that comes from the absence of a cell wall.

The first major morphogenetic movement of vertebrate embryos is gastrulation, the inward movement of cells that is initiated in a particular region of the blastula, the spherical ball of cells that results from cleavage of the fertilized egg (Figure 9.1). Interesting indications about the way in which this may be initiated due to simple biomechanical properties of embryonic cells have been proposed by Mittenthal and Beloussov (1991). Assuming that cells

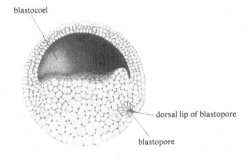

blastocoel

dorsal lip of blastopore

blastopore

Figure 9.1. The initial stage of gastrulation in an amphibian embryo, initiated by the inward movement of cells in a particular reion of the blastula, producing the blastopore.

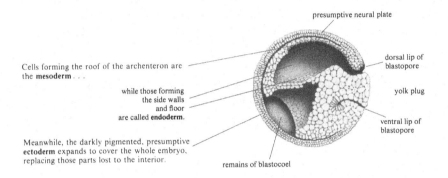

presumptive neural plate

Cells forming the roof of the archenteron are the **mesoderm** . . .

while those forming the side walls and floor are called **endoderm**.

Meanwhile, the darkly pigmented, presumptive **ectoderm** expands to cover the whole embryo, replacing those parts lost to the interior.

dorsal lip of blastopore

yolk plug

ventral lip of blastopore

remains of blastocoel

Figure 9.2. The emergence of a multilayered structure, the gastrula, in which interactions between the primary germ layers generate the structures that define the compex morphology of the emerging embryo.

respond actively to mechanical strains by developing 'hyperrestorative' forces of the type first proposed by Odell et al. (1981), they show how epibolic movements in the blastula (the spreading of the surface cells that starts at the animal pole) generate strains that result in cell deformations and bottle cell formation, initiating the process of invagination as a robust, natural consequence. A blastopore forms from the invaginating cells, which spread over the inner surface of the blastula producing a multilayered structure, the gastrula (Figure 9.2). If the vitelline membrane that normally surrounds the embryo is removed, and the osmotic conditions of the medium in which embryos develop are altered, then instead of invagination followed by migration of cells over the inner surface of the blastula, the blastula buckles out and cells flow over one another to produce an everted sphere, a process called exogastrulation. Because of lack of interaction between outer and inner cell sheets, which normally occurs in the multilayered gastrula and gives rise to further patterns of cell movement and differentiation, the exogastrula fails to develop normally and a nonfunctional, though coherently differentiated, structure is produced. Under normal conditions, the vitelline membrane and osmotic conditions bias the process strongly in favour of invagination and gastrulation. This shows that the mechanical and ionic conditions of the environment are important in determining which of the options (only two in this case) is followed by the developing organism, just as the ionic composition of the medium is important in determining which pattern of growth (local or global deformations) is followed by *Acetabularia*. Organism and environment together define the developmental dynamic and the morphogenetic trajectory.

The next major morphogenetic movement of vertebrate embryos is effectively a repeat of the first, but now the infolding of the cell sheet takes place along a line and forms a tube because of the influence of the inner cell sheet on the outer and the antero–posterior axis that results from gastrulation. The process is called neurulation and the tube formed is the neural tube, from which the nervous system develops. These deformations of cell sheets to produce the processes of gastrulation and neurulation can be simulated by models that treat cells as excitable media of the same kind as that described for *Acetabularia:* the mechanical state of the cytoskeleton alters in interaction with calcium and cells change their shape, giving rise to propagating waves of cell deformation that result in invagination movements. Figure 9.3 shows such a simulation by Odell et al. (1981). A more recent study of notochord formation that accompanies neural tube production, examined in two-dimensional cell sheets, has been presented by Weliky et al. (1991). Other aspects of morphogenesis in animal embryos based upon cytoskeleton–calcium dynamics and its consequences with respect to cell shape change, condensation, and migration have been described by Oster et al. (1985) and by Murray and Oster (1984). The *Acetabularia* model is to be seen as a member of this set.

Near the anterior end of the neural tube, tissue begins to bulge out laterally on both sides and grow into bulbous structures, the optic lobes (Figure 9.4a).These continue to grow laterally until they reach the surface layer of the embryo, the epidermis (Figure 9.4b). On contact of the optic lobe with the epidermis, it flattens to form the optic vesicle and then deforms inward in a repeat of the movement that produces the neural tube, forming the optic cup (Figure 9.4c). As this occurs, epidermal cells respond to the influence of the optic vesicle by undergoing a transition from squamous (flat) to columnar, resulting in a thickening and inward buckling of the sheet (as in gastrulation

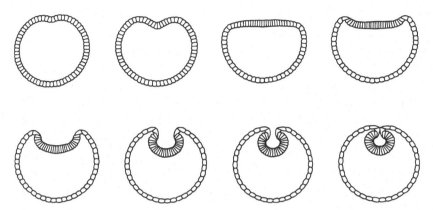

Figure 9.3. The infolding of cells of the neural plate of the gastrula in response to influences from the underlying mesoderm cells that produce the neural tube which will develop into the central nervous system.

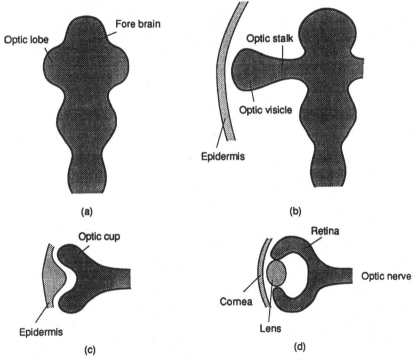

Figure 9.4. The shape changes that occur in the anterior part of the neural tube where the eyes form as a result of outgrowth of the optic lobes, interaction with the epidermis, induction of lens and retina and formation of the optic nerves, which grow back to the midbrain, where they form orderly mappings over the optic tecta.

and neurulation) which eventually results in the detachment of the thickened cells to form the lens (as did the neural plate in forming the neural tube, but now the geometry is circular rather than cylindrical) (Figure 9.4d). The lens becomes transparent, as does the overlying cornea, and the cells of the optic cup differentiate. The outer layer of cells of the retina differentiate into neurons whose axons grow over the surface of the retina and down the optic stalk, forming the optic nerves. These cross over the ventral midline of the embryo and, when they reach the midbrain, they spread out in an organized two-dimensional projection that maps the retina in an ordered manner over the optic tecta of the mesencephelon.

It is now evident that many of the basic morphogenetic events involved in forming the vertebrate eye are repeats of the basic movements that have been encountered over and over again as the natural state changes of morphogenetic fields organised as described in terms of calcium–cytoskeleton dynamics, localised cell growth and deformation, bucklings of cell sheets, changes of cell adhesiveness and directed cell movements over surfaces. These do not account by any means for the full detail of the events that give rise to the highly refined visual systems we see operating in vertebrates. However, that was not the goal, which was the more modest one of suggesting how a primitive but functional system for recording visual images could have arisen independently in many different taxa. It is now clear how simple and natural it is for an embryo to generate a structure of the type shown in Figure 9.4c. With a partially transparent epidermis and excitable cells (neurons) in the optic cup, this already functions as a primitive imaging system, a useful visual organ. This is the first necessary step in the evolution of more sophisticated visual systems, which arise by extensions and refinements of basic morphogenetic movements. The different ways in which vertebrate and invertebrate eyes arise by modifications of these invaginations, foldings and topographic mappings, as well as the basic properties of cells that could give rise to the pigmented, light-sensitive epithelia of retinae, is described in Willmer's (1960) remarkable book dealing with the ways in which cells naturally give rise to complex structures. It seems distinctly possible that the processes involved are robust, high-probability spatial transformations of developing tissues, not highly improbable states that depend upon a precise specification of parameter values (a specific genetic programme).

This conclusion is compatible with the observation that the *eyeless* gene, discovered in Drosophila more than 80 years ago (cf Harland, 1936), has been shown to have homologues in species as diverse as humans, mice, squid, and flatworms, suggesting a common genetic factor for all types of eye despite great differences in structure (Quiring et al, 1994). First, it should be recalled that the *eyeless* gene is not necessary of eye formation in Drosophila: stocks of *eyeless*-flies maintained over several generations recovered eyes despite the continuous absence of a functional *eyeless* gene. And secondly, the *eyeless* gene clearly acts within diverse ontogenetic contexts, and the type of eye that is made is most likely to reflect the robust morphogenetic pathway that leads to an effective visual organ in any particular taxon. The problem remains of accounting for the ontogenetic origins of the different types of eye that have arisen as emergent novelties during the course of evolution. Common genetic ancestry, and the striking capacity of the *eyeless* gene to produce ectopic eyes in Drosophila (Halder et al, 1995) does not eliminate the need to account for the ontogenetic origins of the different morphogenetic pathways used to generate the variety of eyes that have emerged. As Patel (1995) puts it, "evidence of a common ancestry in no way contradicts the view that image-forming eyes evolved independently".

Homology Revisited

There is an interesting embryological detail about vertebrate eye development that provides an opportunity to reiterate the definition of homology that was introduced in Chapter 6, and to relate it to the notion of robust morphogenesis. Lens formation as described earlier in this chapter is a classic example of the inductive effect of one tissue on another. However, there are species of fish, amphibians, birds and mammals in which no induction of head epithelium by the optic vesicle is required for lens formation, prior interaction of epithelium with endoderm and/or mesoderm being sufficient to elicit lens formation (Jacobson and Sater, 1988; Saha, Spann and Grainger, 1989; Grainger, 1992). The detailed tissue interactions and causal events underlying the morphogenetic processes leading to lens formation are therefore not identical in all vertebrates. When faced with this type of diversity of epigenetic mechanism, biologists are tempted to seek phylogenetic patterns and to try to identify the primitive or ancestral condition as in some sense primary. But lens induction by the optic vesicle and the alternatives are apparently distributed without any regularity among the amphibian species that have been studied. Nevertheless, on the basis of cladistic parsimony analysis, the occurrence of an inductive influence from the optic vesicle is regarded as the 'primitive' condition.

These historical questions are certainly of interest in themselves, but they are not relevant to issues of homology as defined in Chapter 6, which is an equivalence relation over a set of structures arising from invariants in the generative process underlying their production. Lens formation in vertebrate species is homologous (equivalently, lenses in vertebrates are homologues) because lenses are generated by morphogenetic processes that have invariant features which are equivalent despite variation of initiating cause and subsequent interactions. These invariant features are the morphogenetic transformations of cell sheets, dependent upon changes of cell shape and cell adhesiveness resulting in detachment of a ball of cells that will form the lens. It is of course always of interest to know the variety of stimuli that can initiate a similar response. Epigenetic processes are notoriously over-determined in this sense, primary embryonic induction being perhaps the most familiar case, in which a great variety of evocating stimuli can initiate the same response. But the definition of homology proposed here is not based upon identity of causal epigenetic detail, nor upon a supposedly "primitive" or ancestral condition, but simply upon the appropriate invariants of the generative process involved, which is generic to the (epigenetic) system and constitutes a robust dynamical process. The usefulness of the homology concept in morphological taxonomies derives precisely from two properties: (1) the generic or typical quality of the generative process, whose invariant features define the equivalence property of the class of structures generated (their homology); (2) its robustness, which underlies the persistence of the generated forms over a diversity of taxonomic groups. As we have seen, homology can be applied at all the different levels of the developmental hierarchy, allowing for comparisons of structures within single organisms, across members of a species and between species at all taxonomic levels. It connects pattern and process by recognising that equivalent patterns arise from generative invariants.

Brian Hall (1991) has discussed these and related issues in considerable detail in his book *Evolutionary Developmental Biology*. The historical dimension of this book gives it some dialectical depth, the central discussion of homology being situated within the terms established by Goethe and elaborated in the great debate between Cuvier and Geoffroy, whose resolution we are still seeking. After an extended discussion of a variety of experimental studies of homology, including tetrapod limbs and vertebrate eyes, Hall finally comes down in favour of a genealogical definition, in basic agreement with Darwin's position. He argues that the epitome of homology is constancy, invariance. Homology of developmental process would therefore require that the entire developmental programs of two parts be *identical* for the two parts to be homologous. Since this is not the case for vertebrate eyes, as we have seen, Hall concludes that homology must be applied to constant final pattern and the historical origins of the pattern, not to generative process, which is variable.

The difficulties with an historical definition of homology were considered in relation to the critique by de Beer (1971, Chapter 6). An appropriate developmental definition of homology therefore revolves about the issue of causal identity versus equivalence of developmental processes involved in the production of similar parts. A cartoon illustrates this issue. Imagine two identical stones rolling down identical slopes, one whose motion was initiated by a push and the other by erosion of retaining soil. These processes are therefore not *identical*. But it is clear that they are dynamically equivalent and can be described as homologous (equivalent under transformation of initiating stimulus). Lens formation with and without induction by the optic vesicle is a similar case. In development, initiating stimulus is frequently a nonspecific trigger that sets off a process that was poised to go, reflecting a state of competence. Waddington (1956, 1957) emphasized this by the use of the term evocator for the stimulus (see also Goodwin, 1991). Homology of developmental processes is thus focused on the identification of intrinsic dynamic equivalences (competences) of generating processes, resulting in characteristic patterns independently of detailed interactions (evocations). It is in fact competence that defines the nature of a process in a generative description, nature being identified here with the intrinsic causal power of a particular morphogenetic field to undergo a particular type of ordered change, as discussed in Chapters 5 and 8.

This discussion allows us to consider another problem of homology, illustrated by the formation of the alimentary canal in vertebrates. In sharks, this forms from the roof of the embryonic gut cavity; in lampreys and newts, it is generated from the floor; in frogs, both floor *and* roof are involved; while in reptiles and birds, it originates from the hypoblast (the lower layer of the embryonic disc). There are embryologists who regard this diversity of tissue origin as evidence against homology of the alimentary canal in the vertebrates, because they wish to identify homology with identity of embryonic origin. This clearly results in problems, since the similarity of the structure in all vertebrates is very obvious and one seeks a definition that accords with this observation. The problems are resolved by the notion of common generative processes involved in gut formation, resulting in structures that constitute an equivalence set over variations (transformations) of tissue origins and causal detail. The invariant generative process is the tendency (the competence) of embryonic cell sheets to form tubes, as already described, different structural details arising in different contexts. This process is generic and robust, resulting in the emergence of typical forms in developing organisms. Alimentary canals in vertebrates are natural manifestations of the same morphogenetic potential whether in the same or in different tissues, from which arises their homology.

The same arguments can be applied to all the major vertebrate structures – eyes, brains, spinal cord and axial skeleton, limbs, hearts and so on – all expressions of the robust, generic properties of the morphogenetic process as it unfolds within a continually transforming context. Major differences of body plan over the taxa – cnidarians, annelids, insects, vertebrates and so on – involve phylotypic stages that define the basic body plans of these groups, which set the contexts within which the morphogenerator unfolds its potential in robust sequences and cascades. But the phylotypic stage (e.g., the insect germ band, the vertebrate pharyngula) is itself an expression of a robust morphogenetic process. The invariant generative events that define these processes involve particular sequences of symmetry breaking that result in typical structures, such as the segmented head, thorax and abdomen of the insect germ band (Sander, 1983) or the brain–spinal cord–segmented axial musculature and basic organ systems of the vertebrate pharyngula (Ballard, 1981). Describing and modelling these basic symmetry-breaking processes is the programme that underlies the construction of a rational taxonomy of biological forms.

The Evolution of Generic Forms

It is useful to examine the evolutionary implications of the proposition that morphogenesis is an intrinsically robust process by first considering its antithesis. It tends to be assumed that what guides morphogenesis through a sequence of stages to a functional adult form is a genetic programme that is stabilized by natural selection – that is, by persistence of those genotypes that generate functionally successful adults in some habitat. The morphological characters of the adult may be very improbable biological structures, but if they contribute significantly to the fitness of the adult, then they will persist as functional components of the species. This conception of an intrinsically improbable structure that has functional value to an organism is described by a sharp peak for the character in a fitness landscape: to produce the character, the genes must be precisely located in a small domain of genotype space, where they are stabilized by natural selection. This implies that there is no need for morphogenesis to be *intrinsically* robust: natural selection can do the stabilizing. Small changes of genotype, or of the environment, can throw morphogenesis off course, but these deviants are then rejected. This is the definition of a low-probability structure: it is but one of a large number of possible forms, all available for evolution, each one reachable from a very small set of initial conditions and requiring a very precise specification of parameters.

The proposition under consideration is just the opposite of this: biological forms are in general the intrinsically robust, high-probability or generic states of the morphogenetic process. They are the 'points of dynamic stability' of morphogenesis. The argument is that natural selection is too weak a force to hold genomes (more strictly, parameter values of developmental dynamics) in small regions of parameter space relative to the changes occurring in fluid genomes, with the consequence that sharp, isolated fitness peaks describing low-probability, functionally useful structures melt under spontaneous genetic variation (Kauffman and Levin, 1987; Kauffman, 1992). Therefore, evolution is necessarily the emergence of robust forms, the generic states of the ontogenetic process. Ontogenesis defines the set of possibilities for phylogenesis.

Genetic Networks and Morphogenesis

In Chapter 8 it was observed that genetic networks behave in evolution as complex dynamic systems capable of a type of learning. Whatever similarities are revealed in the patterns of gene activities generated by these networks during development in a diversity of taxa could arise from historical contingency (common ancestry), or they could reflect the response of an all-purpose learning system to similar constraints encountered by wide classes of developing organism (convergent evolution). The zootype, Geoffroy St Hilaire's 'one animal', could be either the residue of an historical accident or the reflection of similar morphogenetic constraints experienced by plastic genetic networks. I shall now explore an evolutionary scenario in which the latter becomes a plausible conjecture and a model that can assist its systematic exploration.

The basic proposition is that morphogenetic fields, despite their great diversity of expression in the varied adult morphologies of different species, are limited in their possible transformations by intrinsic organisational constraints. This is a proposition as old as 'modern' biology, if by modern we mean starting from Linnaeus and the principles of rational systematics. The position adopted in this volume, as Webster has made clear in Part I, is that the problem of biological form requires a theory of morphogenesis in which the structures of organisms are understood to be a result of dynamic processes that are similar over large taxonomic groups, if not over the total realm of living organisms. It has already been suggested that unicellular and multicellular eukaryotes, plants and animals, may share common morphogenetic processes in the basic dynamics of excitable media, plants and animals differing secondarily in the different constraints imposed by the presence and absence of cell walls. However, excitable media themselves do not define morphogenetic transformations, and so it is necessary to examine the types of constraint that may operate in the animal kingdom, and in particular throughout the metazoa, that could provide a foundation for explaining the morphospace of possibilities available to this extensive and varied taxon. Clearly at this preliminary stage in the construction of a realist theory of the natural kinds of animals, it is possible to do no more than point in a certain direction by using a particular model that has deep resonances with the empirical evidence at one level of the taxonomic hierarchy.

The elements required of a morphogenetic theory of the metazoa include (1) an explicit description of a nonlinear dynamic morphogenetic field and how its states determine particular three-dimensional forms; (2) the dynamics of genetic networks which allow them to explore the possibilities available to the morphogenetic field and to stabilize 'successful' morphologies.

A recent proposal that allows us to examine the type of model required by (1) can be found in a paper by Cummings (1994). In this he defines a general class of nonlinear field equations that generate spatial patterns of 'morphogens' within an initially defined spatial domain. These morphogens are not necessarily of Turing type described by biochemical reactions and diffusion of products. Rather, they can be gene products that are generated locally by a nonlinear reaction system of the general type considered in Chapter 7 in relation to gene regulators, including transcription factors. However, the particular gene products designated 'morphogens' by Cummings are assumed to have an influence on cell shape via such processes as cell–cell adhesion, change of cell shape, cell growth and division and so forth. This influence is described specifically in terms of curvature.

Curvature can be positive or negative on a particular surface depending on which way the surface curves, whether 'inwards' or 'outwards'. These relative terms depend upon selection of a standard direction for measuring the radius of curvature. Wherever there is a saddlelike region of the surface, the curvature will be positive along one direction (say the line traced out by your legs as you sit in the saddle) and negative in the other direction (the saddle curving up in front of you and behind). The radii of curvature along these principal directions extend from the surface in opposite directions, down and up. When a shape such as a blastula (Figure 9.1) transforms into a gastrula (Figure 9.2), a surface of positive curvature (a sphere) transforms through the development of regions of negative curvature (where the surface buckles in and invagination occurs) to one in which there are saddlelike regions around the lips of the blastopore. Here the surface curves locally in different directions along lines from outside to inside the blastopore and along lines running around the blastopore.

Cummings (1994) describes metazoan form in terms of the curvatures of sheets of cells which can change shape in particular ways in response to local concentrations of 'morphogens'. His cells are elements of structure defining the deformable sheets, and they need not correspond precisely to biological cells. He imposes realistic constraints on how these 'cells' can change their shapes, from regular cuboids to elements contracted on one side and expanded on the other, including expansion and contraction along different axes so that saddlelike deformations can occur. From these allowed local transformations he describes the changes of curvature that can result in the whole surface. He shows that the constraints allow the sheets of cells to transform in all the characteristic ways that occur during metazoan development: gastrulation, neural tube (cylinder) formation (Figure 9.3), buckling out or in of the surface, pinching off structures, periodicities of structure as in segmentation and so on. The particular transformations that the system undergoes depend upon local morphogen values which determine cell shape and hence local curvature, involving a single parameter. This parameter is assumed to depend upon local morphogen values. These morphogens (only two in Cummings's model) have values determined by morphogenetic field equations which, as described earlier, are like Turing equations but are actually more comprehensive and do not depend upon a particular mechanism such as reaction and diffusion. These equations also have parameters which determine their detailed behaviour, and these are assumed to be dependent upon gene activity.

Here is a model that goes from gene activity via a morphogenetic field to the precise specification of three-dimensional form, the morphology of the developing organism. As gene activity changes, morphogenesis occurs and the organism changes its form. However, a series of morphogenetic transformations will occur even if the only activity of the genes is to allow growth to occur, or some process equivalent to growth in terms of the model (e.g., change of cell size or modification of the effective communication distance between cells). That is to say, genes don't have to do very much to produce morphogenesis in this model because of the coupling between morphogens and curvature on the one hand, and reciprocal effect of curvature (shape) and 'size' on the morphogenetic field solutions.

This mutual coupling between the dynamics of a nonlinear morphogenetic field (with bifurcations and symmetry breaking) and geometry or morphology, is what emerged as a crucial feature of the analysis of the *Acetabularia* model described in the previous chapter. There growth was a primary component of the model. All the parameters ('genes') could remain constant, providing growth occurred and the parameter values were located in a particular region of parameter space. What Cummings's model provides is a mathematically more elegant and versatile description of this type of process, together with a variety of other transformations that are characteristic of the metazoa. It provides a foundation for the first property required of a morphogenetic theory as described in condition (1) on p. 245.

The second requirement is a dynamic process whereby the morphospace of possible forms can be explored by the system itself. This requires genetic networks within the 'cells' of the organism that can modify parameters and thus alter morphogenetic sequences. It also requires that the states of the genetic networks be affected by the local values of the morphogenetic field. One way of achieving this is to have gene activity affected by local curvature. The implication is that gene activity is influenced by variables that depend upon mechanical strain in cells, such as substances whose intracellular levels depend upon strain-dependent membrane pumps and channels, or by the state of the cytoskeleton and the nuclear skeleton influencing gene activities via modification of translation and transcription rates (see, e.g., Stein and Bronner, 1989). However, what is important in the model is not so much the precise details of the coupling as the logical structure of the whole system. To generate a space of possible forms (a morphospace) the system must be autonomous or logically closed in the following sense (assuming a constant environment for the moment): the states of the genetic network control parameters that specify the morphogen patterns and hence the curvature, local state of the 'cells' influencing the states of genes in the network. This type of system has the property that nothing is in overall control because the cycle of influences is dynamically closed. We need now to expand on the structure of the genetic network as a learning system within this context.

The genetic networks require two basic properties to assist in the exploration of morphospace. There must be a source of variation of the 'genes' and their interactions, which can be either random or directed (to be discussed shortly). And the networks must have the capacity to stabilize dynamic patterns which result in parameter changes that produce life-cycle closure. By this I mean that the 'organism' starts with a particular shape (say a hollow sphere of cells), proceeds through a sequence of morphogenetic transformations and generates from itself more of the initial reproductive structures (spheres of cells) which then repeat the cycle. Such sequences represent asexual reproduction by a budding process, as in zooid formation in coelenterates (cnidarians and ctenophores), tunicates and other phyla. Morphogenetic sequences of this type can be generated in Cummings's model as the morphogenetic field passes through a bifurcation sequence and local morphogen values induce shape changes via their effects on local curvature.

This pattern of life-cycle closure can be taken to be the simplest criterion of natural selection or dynamic stability of 'successful' morphological reproduction. For this to occur, it is necessary for the spherical 'bud' that is generated by the adult form in the model to consist of cells whose genetic networks are in the same (or sufficiently similar) state as those in the initiating sphere of cells that gave rise to the adult form, so that the life cycle will repeat. The genetic networks must learn to stabilise such closed life cycles. Since closed life cycles are the only ones that result in reproduction, genetic networks that enhance the stability of such cycles to random fluctuations will result in increased abundance of the 'species', while networks that reduce the stability will result in decreased abundance. Therefore it can be expected that the former will increase relative to the latter.

Because of nonlinearities, similar morphogenetic sequences could be generated by rather different genetic network sequences, and life-cycle closure by different cycles of the networks. A primary question then is: what effects do the limited set of transformations of a morphogenetic model such as Cummings's have on the patterns of gene activity that result in successful life cycles? Will the constraints of such a model result in constrained sequences and spatial patterns of change in the genetic networks of evolved 'species' (defined by morphology in this case) that result in a conserved zootype? It is generally assumed that phenotypes are extremely plastic and genotypes are prone to historical constraints. However, constrained morphogenetic transformations may be the origin of both the logical order of taxonomic relationships and common patterns of gene expression across extensive phyletic domains. The evidence for a polyphyletic origin of large taxa such as coelomates and arthropods, and other patterns of convergent evolution as described in Minelli (1993), point to morphogenetic constraints on evolutionary transformation, but of course these questions of polyphyly versus monophyly are under active discussion and there is as yet no consensus within the cladistic community. Having a model with which to explore evolutionary scenarios with plausible morphogenetic life cycles could help to clarify these issues. Although the model sketched here is initially restricted to asexual reproduction, this is not a bad place to start in the construction of a more general theory that is capable of indefinite, though not arbitrary, modification in the direction of more realistic and comprehensive life cycles and even ecosystems (extending selection or life-cycle stability criteria to include interactions with other organisms). The virtue of the approach is that morphogenesis is included in the modelling exercise, with the result that organisms return to biology as essential causal agents. This model is now under investigation.

Generative Dynamics Includes the Environment

At this point an extension of the discussion, which has been focussed largely on organisms and their parts, is required in order to be more explicit about the relevance of the environment in developmental processes. As described explicitly in relation to *Acetabularia* morphogenesis, the environment not only defines essential parameter values such as the ambient calcium concentration; it is also involved in the developmental dynamics of this species by acting as a conducting medium that makes possible the occurrence of electrical currents that are an essential accompaniment of growth and morphogenesis. This phenomenon is very widespread in developing systems, as emphasised by Jaffe (1981), though in cases like human fingertip regeneration the external sea water is replaced by body fluids that provide the extracellular ionic conductor, the original environment being replaced by a medium that has a physiological origin and we can speak of the original external environment being internalized. This is a common evolutionary strategy. However, the external environment is always involved in defining the conditions of ontogenesis and participates in this process. The robust or generic states of the life cycle are then those defined by this extended dynamical system in which the organism (both developing and adult) together with its environment (including other species) constitute a single unified process. The organism is influenced by, and influences, its habitat. This dialectical unity of organism and environment has been eloquently described by a number of writers on evolution, among them Waddington (1957), Ho and Saunders (1979), Levins and Lewontin (1985) and Oyama (1985, 1989); the dynamical process whose attractors define species life cycles includes relevant environmental variables and parameters as components of the system. The 'external' and the 'internal' are inexorably folded together: the chloride and calcium ions that flow through the cytoplasm of a developing *Acetabularia* cell are continuous with those flowing outside, defining a single current loop. Patterns of species interactions involving both cooperation (symbiosis, mutualism, etc.) and competition are included in the dynamic. So the morphogenetic field becomes a unitary dynamic process, still a field since space and time are involved in the generative dynamics of the life cycle defined by characteristic morphology and behaviour (form), but now extended to explicitly include relevant environmental components.

The proper analytical context for the integration of development and evolution that is being proposed here is the theory of complex dynamical systems, species life cycles being the attractors of the dynamic. Figure 8.14 can be used to make this proposal more explicit. There the concept of a generic form was used to describe the high-level taxonomic signature of an order, the *Dasycladales*. This generic form can be regarded, from the point of view of the developmental trajectories that start from within the invariant set of parameters (genes and environmental factors), as a robust attracting region of the whole set of life cycles that collectively define the order of giant unicellular green algae. Each species is described by a closed life cycle from a parameter set that defines the species, through the generic form of the species and back, via gametogenesis, to the initial parameter set. The species life cycle is a dynamic attractor whose characterisation remains to be specified, whether a limit cycle with stochastic perturbations or more like a strange attractor with intrinsic unpredictability of successive trajectories that nevertheless remain confined within a domain, that of species viability. What is called natural selection is then described by the stability of these species life cycles, which include the full range of ecosystem interactions on which the species depends. But species morphologies are now seen as the result of robust morphogenetic trajectories. These forms are not historical individuals resulting from the accidents and contingencies of evolutionary history. They are natural kinds, which within a realist philosophy are classes of things produced by common generative mechanisms (Bhaskar, 1978; see Chapter 5). The different forms have relationships of similarity and difference that depend upon the intrinsic dynamics of the generative processes that produce these forms – that is, relationships dependent upon the sequence of symmetry-breaking ontogenetic cascades whereby biological forms arise. A primary task of biology as a realist science is to classify these forms (species) in terms of their internal generative relationships, thus producing a rational taxomony of species. Such a taxonomy is not dependent on history, just as the periodic table of the elements, the physicist's rational taxonomy of physical forms, is history independent, though of course the actual generation of members of this set occurs under particular conditions at particular times in cosmic evolution. History is always involved in any generative process. The question at issue is whether or not it explains the relationships of similarity and difference that are recognized in homology and taxonomy.

Random and Directed Mutations: Neo-Darwinism *and* Lamarckism

The evolutionary process starts in a particular domain of its dynamical space and proceeds by an exploration of all possible variants of this complex system by parametric variation, by change of dimensionality (introduction of new variables and degrees of freedom), change in the number of levels of the developmental hierarchy and the degree of modularity at each level, and so on. These variations, arising primarily from genetic and environmental change, can be either random, as assumed in conventional neo-Darwinism, or 'directed', as described in the studies of Cairns et al. (1988) and Hall (1988, 1990) on bacterial adaptation. These authors interpreted their results as evidence that individual organisms are capable of altering their genomes adaptively in response to particular environments and passing them on to their offspring, which is a Lamarckian pattern of inheritance. As expected, there is strong resistance to this interpretation, which breaches Weismann's barrier (cf. Lenski and Mittler, 1993). Crucial to this dispute is the question of just how labile the genetic material is, and how its stability changes with growth conditions. It is now generally recognised that the DNA is as fluid and changeable as the rest of the organism (cf. Ho, 1986). The issues raised by Lenski and Mittler concern whether or not the molecular transformations and transpositions of DNA result in the capacity of organisms to respond in a directed manner to environmental conditions. An assessment of this controversy that focusses on the very different cell states under conditions allowing growth (conventional laboratory selective environments) and those that do not (starvation or basic maintenance conditions, as used by Cairns, Overbaugh and Miller 1988) has led Symonds (1994) to the conclusion that "there can be no question that an element of directed mutagenesis will be found amongst stationary cultures that are maintained under selective pressure. The degree of directionality that applies in any particular case is a question for experiment, as is the crucial task of determining what genetic and biochemical pathways are involved." This view is now receiving substantial support that seems to be leading towards a consensus (Rosenberg et al., 1994; Foster and Trimarchi, 1994; Jablonka and Lamb, 1995).

The door is thus opened to a reassessment of many questions relating to rates and directions of evolution under different conditions. It also emphasises the dynamic unity of the whole organism, without privileged parts such as its DNA, and the dynamic coupling of the whole living process with its environment. This extends now to the role of life in altering geophysical processes, as emphasised particularly in the Gaia hypothesis (Lovelock 1979, 1991). Dynamical analysis is no respecter of disciplines whether it be development, genetics, ecology, geology or other, and all components relevant to a complete description of the process under study must be included if the analysis is to have validity. The process studied may well split up into hierarchical levels that to some extent reflect traditional disciplinary boundaries, but the challenge now is to leave these divisions behind in integrated studies of complex systems that provide more holistic perspectives on natural processes than traditional disciplinary boundaries have allowed.

The recognition that both neo-Darwinian *and* Lamarckian mechanisms of heritable variation are involved in evolutionary transformations broadens significantly the dynamical basis for the emergence of coherent novelty in organismic morphology and behaviour. Genetic networks can undergo extensive reorganisation as the whole organism responds to challenges and opportunities of its own and the environment's making. This provides for a continuous exploration of the generative fields of living forms, which emerge as dynamic attractors. The evolving system falls into basins of attraction defining species life cycles, which basins can be large or small, deep or shallow or anything in between. The stability of a species is then determined by the characteristics of the basin of attraction that defines the life cycle: small basins will tend to be unstable to parametric change, while large basins will tend to be stable.

I have been focussing mainly on the large, deep basins that characterise the stable morphological taxa on which classification depends – the hierarchy of classes defined by transformations over equivalence sets such as segments in invertebrates, limbs in vertebrates, leaf and flower patterns in plants. However, as well as these robust forms there are many others of varying degrees of stability up to the highly transient, labile forms that cause such taxonomic headaches. All of these morphological classes have norms of reaction that characterise the stability of the attractor. These norms define the robustness of the developmental trajectories to genetic and environmental perturbation and are embodied in Waddington's (1957) notion of canalisation, the degree of dynamic stability of an epigenetic pathway which can also change in time. And this whole evolutionary unfolding is intrinsically unpredictable, a property that is characteristic of complex dynamic systems. A rational taxonomy of biological forms is necessarily open to unexpected emergent novelties, which are nevertheless intelligible in retrospect.

Rational Taxonomy

This view of species as natural kinds implies, as previously noted, the existence of a rational taxonomy of forms, a classification of the attractors according to the intrinsic properties of the dynamics. Hierarchical dynamics results in a hierarchical taxonomy of morphologies, as observed in Chapter 6. An example illuminating this intimate connection between ontogenetic and taxonomic hierarchies is provided by a classification of teratologies in *Drosophila* by Ho (1990). Exposure of embryos to ether vapour for a brief ($\sim$ 20 minutes) period during the first 3 hours of development results in a characteristic spectrum of segmentation abnormalities in unhatched and hatched larvae (Ho et al., 1987). Using generative rules that are based upon the sequence of global spatial harmonics of decreasing wavelength on which different categories of segmentation gene are expressed during this period (described in Chapter 7 and in Goodwin and Kauffman, 1990), Ho predicted a set of possible disturbed segmentation categories that were much fewer than expected if each segment was independently perturbable. The constrained set of possibilities arises from the assumption that each level of the segmentation hierarchy must be established before the next lower (shorter wavelength) pattern of spatial order and gene influence can be successfully expressed (see Figure 9.5). This defines an invariant generative rule of seg-

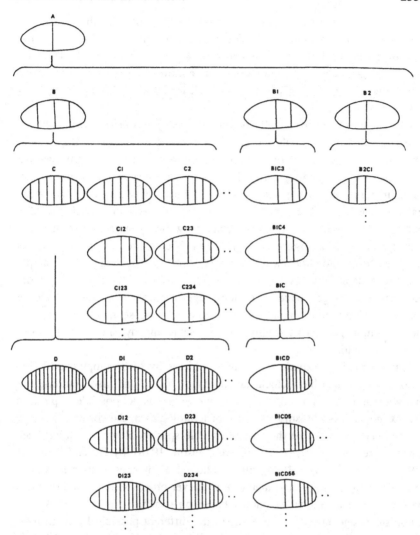

Figure 9.5. The hierarchical sequence of segmentation events shown as successive subdivisions of the embryo corresponding to the action of maternal genes (A), gap genes (B), pair-rules (C) and segment polarity genes (D), together with possible mutant phenotypes. B_1 means that the left-hand subdivision at step B failed to occur, with the consequence that all subsequent subdivisions dependent on B_1 also fail. Similarly for other mutant phenotypes, some of which correspond to genetic mutants, as shown.

mentation. The robust morphogenetic property that underlies this generative process is the development of spatial periodicities and their harmonic progression through a bifurcation sequence of decreasing wavelength. The possible forms generated can be organized into a rational taxonomy of segmental teratologies in which neighbouring forms differ by a single step in the generative process.

When a set of experimentally perturbed *Drosophila* larvae was analysed, it was found that they could be identified with the categories of the rational taxonomy. Using a set of twenty-four larvae with a range of perturbations, Ho constructed an ontogenetic cladogram that locates the segmentation forms in terms of 'ontogenetic distance', the number of generative steps that separate them. The result is a hierarchical taxonomy consisting of a single 'order' with two 'families' containing two genera in the first and eight in the second (Figure 9.6). Attempts to construct a consensus parsimony cladogram on the basis of the segmental characters of the sample failed to generate the 'true' ontogenetic cladogram, which is not surprising since the latter is generated by a knowledge of ontogenetic rules which are not revealed by final characters. The homology of segments arises from their common generative origins in relation to the dynamics of the morphogenetic field.

This instructive example shows how a hierarchy of developmental bifurcations results in a hierarchical taxonomy of forms, and also demonstrates that without a knowledge of the generative rules it is not possible to produce the taxonomy. A criticism of the use of teratologies in the construction of a rational taxonomy is that they are not viable species. However, as has been argued here and elsewhere, notably by Nelson (1978), by Ho and Saunders (1979) and by Webster (1984), all morphological changes in species are due to ontogenetic changes, and therefore whatever order exists in the latter will be revealed in a rational taxonomy based on generative principles. A definition of ontogenetic distance between the different possible forms in morphospace requires a generative theory of the general type provided by Ho (1990) in order to identify the number of generative steps that separate any two forms, providing a measure of taxonomic distance.

When taxonomy is treated as genealogy, then the logical relationship between ontogenetic pathway and taxonomic distance is lost, because neighbouring species are then determined by historical sequence rather than by generative relationships. The historical succession of species has no obligation to proceed systematically through the space of possible forms. Though there will be a tendency for the neighbourhood relations of attractors to be expressed in the sequence of species discovered by evolution, parametric

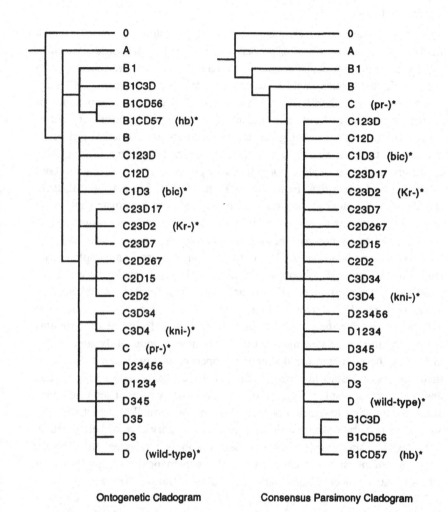

Ontogenetic Cladogram   Consensus Parsimony Cladogram

Figure 9.6. The ontogenetic cladogram describes the neighbourhood relations of the different phenotypes of the sample used in terms of ontogenetic distance, measured as the number of segmentation events that separate the phenotypes according to the successive subdivision model. Attempts to resolve the set of phenotypes by using a consensus parsimony cladogram failed to generate the ontogenetic cladogram.

variation can be discontinuous and jump from one domain of parameter space to another, missing intervening attractors, which may be encountered later. So genealogy, historical taxonomy à la Darwin, will in general not be the same as a rational taxonomy.

History and Logic

Nevertheless history has a distinctive role to play in biological process which arises from the basic dynamic property of life cycles: the initial state of a developing organism depends upon parameter values that are repeated with every cycle that remains within the attractor defining the species. This cycle is the process of inheritance which involves DNA, the developmental or epigenetic system, and the environment. It is the dynamic closure (or quasi-closure) of this process that gives to organisms their distinctive properties and allows them to step into spaces that inanimate systems lacking such cycles are unable to explore. Organisms as physical systems obey the laws of physics and chemistry. However, as Pichot (1994) has aptly put it, there is a physical-chemical gap between the living state and inanimate nature that, in general, increases as evolution proceeds. Initially, with the origin of life, this gap is virtually nonexistent and the reproductive cycle of primitive 'living' systems can be understood in terms similar to those of a complex phys-ico-chemical reaction system undergoing a dynamic cycle in which its constituents are reproduced (Kauffman, 1986; Bagley, Farmer and Fontana 1992; Fontana, 1992). However, as these closed cycles undergo heritable variation they have the possibility of moving further from the states readily reached by inanimate systems and arriving at new patterns of dynamic order. Gradually there emerge the distinctive properties of organisms as we know them, agents or 'things-in-process' (Emmet, 1992) that occupy dynamic states and attractors far displaced from those of inanimate systems. They reach these states by coupling together processes that do not normally interact strongly, such as concentrations of particular proteins and curvature of cell sheets as described in Cummings's (1994) model. Yet these processes still obey the principles of physics and chemistry, with mathematical descriptions of the emergent states of novel order and their rules of transformation.

The study of biological form explored in this book involves attempts to identify these rules and so to make sense of the relationships between organisms of different type in terms of equivalence classes in the space of generative dynamics (fields) which define the causes of organismic morphology and behaviour. It is by no means clear that such a programme will succeed. However, if biology is ever to transcend the limitations of an historical science and join history to generative logic, a further attempt must be made along the path already defined by the rationalist tradition. The growing community of inquirers with this orientation suggests that the time is now ripe for another expansion of this frontier.

References

Akam, M. (1987). The molecular basis for metameric pattern in the *Drosophila* embryo. *Development 101*, 1–22.

Atran, S. (1990). *Cognitive Foundations of Natural History: Towards an Anthropology of Science.* Cambridge: Cambridge University Press.

Bagley, R. J., Farmer, J. D., and Fontana, W. (1992). Evolution of a metabolism. In *Artificial Life II*, ed. C. G. Langton, C. Taylor, J. D. Farmer and S. Rasmussen, pp. 141–158, vol X, Santa Fe Institute Studies in the Sciences of Complexity. Reading, Mass.: Addison-Wesley.

Balcuns, A., Gasseling, M. T., and Saunders, J. W., Jr. (1970). Spatio-temporal distribution of a zone that controls anteroposterior polarity in the limb bud of the chick and other bird embryos. *Amer. Zool. 10*, 323.

Ballard, W. W. (1981). Morphogenetic movements and fate maps in vertebrates. *Amer. Zool. 21*, 391–399.

Barinaga, M. (1995). Focusing on the *eyeless* gene. Research News Report. *Science 267*, 1766–1767 (quote from N. H. Patel).

Bateson, B. (1928). *William Bateson, F. R. S., Naturalist.* Cambridge: Cambridge University Press.

Bateson, W. (1886). The ancestry of the Chordata. *Quart. J. Micros. Sci. 26*, 535–571.

(1892). On numerical variation in teeth, with a discussion of the conception of homology. *Proc. Zool. Soc. London*, 102–115.

(1894). *Materials for the Study of Variation.* London: Macmillan. Reprinted Baltimore: Johns Hopkins University Press, 1992.

(1913). *Problems of Genetics.* New Haven, CT: Yale University Press.

Baumgartner, S., & Noll, M. (1991). Network of interaction among pair-rule genes. *Mech. Dev. 33*, 1–18.

Berger, S., & Kaever, M. J. (1992). *Dasycladales: An illustrated Monograph of a Fascinating Algal Order.* Thieme: Stuttgart.

Berrill, N. J. (1952). Regeneration and budding in worms. *Biol. Rev. 27*, 401–438.

(1971). *Developmental Biology.* New York: McGraw-Hill.

Bertalanffy, L. von, and Woodger, J. H. (1933). *Modern Theories of Development.* Oxford: Oxford University Press.

Bhaskar, R. (1978). *A Realist Theory of Science.* Brighton: Harvester.

Blanchard, J. (1990). *Narcissus: A Guide to Wild Daffodils.* Woking, Surrey: Alpine Garden Society.

Bowles, E., and Allen, N. S. (1986). A vibrating probe analysis of light dependent transcellular currents in *Acetabularia*. In *Ionic Currents in Development*, ed. R. Nuccitelli, pp. 113–121. New York: Alan R. Liss, USA.

Brady, R. H. (1987). Form and cause in Goethe's morphology. In *Goethe and the Sciences: A Re-appraisal*, ed. F. Amrine, F. J. Zucker and H. Wheeler. Dordrecht: Reidel.

Brandts, W. A. M., and Trainor, L. E. H. (1990a). Non-linear field model of pattern formation: intercalation in morphalactic regulation. *J. Theoret. Biol. 146*, 37–56.

(1990b). Non-linear field model of pattern formation: application to intracellular pattern reversal in *Tetrahymena*. *J. Theoret. Biol.* 146, 57–86.

Brière, C., and Goodwin, B. C. (1988). Geometry and dynamics of tip morphogenesis in *Acetabularia*. *J. Theoret. Biol. 131*, 461–475.

(1990). Effects of calcium input/output on the stability of a system for calcium-regulated viscoelastic strain fields. *J. Math. Biol. 28*, 585–593.

Brooks, W. K. (1883). *The Law of Heredity: A Study of the Cause of Variation and the Origin of Living Organisms*. Baltimore: John Murphy.

Buck, R. C., and Hull, D. L. (1966). The logical structure of the Linnaean hierarchy. *Systematic Zool. 15*, 97–111.

Burstein, Z. (1995). A network model of developmental gene hierarchy. *J. Theoret. Biol. 174*, 1–12.

Buss, L. W., and Dick, M. (1992). The middle ground of biology: themes in the evolution of development. In *Molds, Molecules, and Metazoa: Growing Points in Evolutionary Biology*, ed. P. R. Grant and H. S. Horn, pp. 77–97. Princeton, N.J.: Princeton University Press.

Cairns, J., Overbaugh, J., and Miller, S. (1988). The origin of mutants. *Nature 335*, /001014/145.

Cameron, J. A., and Fallon, J. F. (1977). Evidence for a polarizing zone in the limb buds of *Xenopus laevis*. *Dev. Biol. 55*, 320–330.

Carr, B. (1987). *Metaphysics: An Introduction*. London: Macmillan.

Carroll, S. B., and Scott, M. P. (1986). Zygotically active genes that affect the spatial expression of the *fushi tarazu* segmentation gene during early *Drosophila* embryogenesis. *Cell 45*, 113–126.

Carroll, S. B., and Vavra, S. H. (1989). The zygotic control of *Drosophila* pair–rule gene expression. II. Spatial repression by gap and pair-rule gene products. *Development 107*, 673–683.

Cassirer, E. (1923). *Substance and Function*. Reprinted New York: Dover, 1953.

(1944). The concept of group and the theory of perception. *Philosophy and Phenomenological Research 5*, 1–35.

(1950). *The Problem of Knowledge*. New Haven, Conn.: Yale University Press.

(1957). *The Philosophy of Symbolic Forms:* vol. III, *The Phenomenology of Knowledge*. New Haven, Conn.: Yale University Press.

(1979). Reflections on the concept of group and the theory of perception (1945). In *Symbol, Myth, and Culture. Essays and Lectures of Ernst Cassirer 1939–1945*, ed. D. P. Verene. New Haven, Conn.: Yale University Press.

(1981). *Kant's Life and Thought*. New Haven, Conn.: Yale University Press.

Caws, P. (1988). *Structuralism: The Art of the Intelligible*. Atlantic Highlands, N.J.: Humanities Press International.

Chasan, R., Jin, Y., and Anderson, K. V. (1992). Activation of the *easter* zymogen is regulated by five other genes to define dorsal–ventral polarity in the *Drosophila* embryo. *Development 115*, 607–616.

Churchill, F. B. (1969). From machine theory to entelechy: two studies in developmental teleology. *J. Hist. Biol. 2*, 165–186.

Cleland, R. (1971). Cell wall extension. *Ann. Rev. Plant Physiol.* 22, 197–222.

Coates, M. I. (1994). The origin of vertebrate limbs. In *The Evolution of Developmental Mechanisms. Development* 1994 Supplement, 169–180.

Coates, M. I., and Clack, J. A. (1990). Polydactyly in the earliest known tetrapod limbs. *Nature 347*, 66–69.

Cohen, J., and Stewart, I. (1994). *The Collapse of Chaos. Discovering Simplicity in a Complex World.* London: Viking.

Coulter, D. E. R., and Wieschaus, E. (1988). Gene activities and segmental patterning in *Drosophila:* analysis of odd-skipped and pair-rule double mutants. *Genes and Dev.* 2, 1812–1823.

Cummings, F. W. (1994). Aspects of growth and form. *Physica D 79*, 149–163.

Darwin, C. (1859). *On the Origin of Species by Means of Natural Selection.* Reprinted Harmondsworth: Penguin Books, 1968.

(1889). *The Descent of Man.* 2nd edn. London: John Murray.

Dawkins, R. (1976). *The Selfish Gene.* Oxford: Oxford University Press.

(1986). *The Blind Watchmaker.* London: Longmans.

de Beer, G. R. (1932). *Vertebrate Zoology.* Oxford: Oxford University Press.

de Beer, Sir Gavin (1971). *Homology: An Unsolved Problem.* Oxford: Oxford University Press.

Depew, D. J., and Weber, B. H. (1995). *Darwinism Evolving. System Dynamics and the Genealogy of Natural Selection.* Cambridge, Mass.: MIT Press.

de Vries, H. (1900). The law of segregation of hybrids, Reprinted in C. Stern and E. R. Sherwood, *The Origin of Genetics.* San Francisco: Freeman, 1966.

Delisi, C. (1988). The Human Genome Project. *Amer. Sci. 76*, 488–493.

Desmond, A., and Moore, J. (1991). *Darwin.* London: Michael Joseph.

Douady, S., and Couder, Y. (1992). Phyllotaxis as a physical self-organised growth process. *Phys. Rev. Lett. 68*, 2098–2101.

Driesch, H. (1908). *The Science and Philosophy of the Organism.* London: Black.

(1914). *The History and Theory of Vitalism.* London: Macmillan.

Driever, W., and Nüsslein-Volhard, C. (1988). A gradient of *bicoid* protein in *Drosophila* embryos. *Cell 54*, 83–93.

Dupré, J. (1981). Natural kinds and biological taxa. *The Philosophical Review 90*, 66–90.

Emmet, D. (1984). *The Effectiveness of Causes.* London: Macmillan.

(1992). *The Passage of Nature.* London: Macmillan.

Fallon, J. F., and Crosby, G. M. (1977). Polarizing zone activity in limb bud amniotes. In *Vertebrate Limb* and *Somite Morphogenesis*, ed. D. A. Ede, J. R. Hinchliffe, & M. Balls, pp. 55–69. Cambridge: Cambridge University Press.

Faraday, M. (1859). *Experimental Researches in Electricity*, vol. II, p. 284.

Feigenbaum, M. J. (1978). Quantitative universality for a class of non-linear transformations. *J. Stat. Phys. 19*, 25–52.

Fisher, R. A. (1936). Has Mendel's work been rediscovered? Reprinted in C. Stern and E. R. Sherwood, *The Origin of Genetics.* San Francisco: Freeman, 1966.

Fontana, W. (1992) Algorithmic chemistry. In *Artificial Life II*, ed. C. G. Langton, C. Taylor, J. D. Farmer and S. Rasmussen, pp 159–209, vol X, Santa Fe Institute Studies in the Sciences of Complexity. Reading, Mass. Addison-Wesley.

Foster, P. L., and Trimarchi, J. M. (1994). Adaptive reversion of a frameshift mutation in *Escherichia coli* by simple base deletions in homopolymeric runs. *Science 265*, 408–410.

Foucault, M. (1970). *The Order of Things.* London: Tavistock.

Frankel, J. (1979). An analysis of cell surface patterning in *Tetrahymena.* In *Deter-*

minants of Spatial Organization, Soc. Dev. Biol. Symp. 37, ed. S. Subtelny and I. R. Konigsberg, pp. 215–246. New York: Academic Press.

(1989). *Pattern Formation.* Oxford: Oxford University Press.

Frankel, J., and Jenkins, L. M. (1979). A mutant of *Tetrahymena thermophila* with a partial minor-image duplication of cell surface pattern. II. Nature of genic control *J. Embryol. Exp. Morphol.* 49, 203–227.

Frankel, J., Jenkins, L. M., Bakowska, J., and Nelson, E. M. (1984). Mutational analysis of the patterning of oval structures in *Tetrahymena*. I. Effects of increased size on organization. *J. Embryol. Exp. Morphol.* 82, 41–66.

Frankel, J., Jenkins, L. M., Nelson, E. M., and Stoltzman, C. A. (1993). *hypoangular:* a gene potentially involved in specifying positional information in a ciliate, *Tetrahymena thermophila. Dev. Biol.* 160, 333–354.

Frankel, J., Nelson, E. M., Bakowska, J., and Jenkins, L. M. (1984). Mutational analysis of paterning in oval structures in *Tetrahymena*. II. A graded basis for the individuality of intracellular structural arrays. *J. Embryol. Exp. Morphol.* 82, 67–95.

French, V. (1993). The long and the short of it. *Nature 361,* 400–401.

French, V, Bryant, P. J., and Bryant, S. V. (1976). Pattern regulation in empimorphic fields. *Science 193,* 969–981.

Galton, F. (1889). *Natural Inheritance.* London: Macmillan.

Gaul, U., and Jäckle, H. (1989). Analysis of maternal effect combinations elucidates regulation and function of the overlap of *hunchback* and *Krüppel* gene expression in the *Drosophila* blastoderm embryo. *Development 107,* 651–662.

Gergen, J. P., and Wieschaus, E. F. (1985). The localized requirements for a gene affecting segmentation in *Drosophila:* Analysis of larvae mosaic for *runt. Dev. Biol. 109,* 321–335.

Gergen, J. P., Coulter, D., and Wieschaus, E. (1986). Segmental pattern and blastoderm cell identities. In *Gametogenesis and the Early Embryo,* pp. 195–200. New York: Alan R. Liss.

Ghiselin, M. T. (1974). A radical solution to the species problem. *Syst. Zool. 23,* 536–544.

Giacomoni, P. (1993). *Le Forme e il Vivente: Morfologia e Filosofia della Natura in J. W. Goethe.* Naples: Guida Editori.

Glass, L. (1977). Patterns of supernumerary limb regeneration. *Science 198,* 321–322.

Glass, L., and Mackey, M. C. (1988). *From Clocks to Chaos.* Princeton, N.J.: Princeton University Press.

Goethe, J. W. (1970). *Italian Journey [1786–1788].* Harmondsworth: Penguin Books.

Goethe, J. W. von. (1946) trans. A. Arber. In Goethe's Botany, *Chronica botanica.* 10, 63–126.

Goldschmidt, R. B. (1945). Additional data on phenocopies and gene action. *J. Exp. Zool. 100,* 193–201.

(1958). *Theoretical Genetics.* Berkeley: University of California Press.

Goodwin, B. C. (1963). *Temporal Organization in Cells.* New York: Academic Press.

(1976). *Analytical Physiology of Cells and Developing Organisms.* Academic Press.

(1982). Development and evolution. *J. Theoret. Biol. 97,* 43–55.

(1990). Structuralism in biology. *Science Progress Oxford 74,* 227–244.

editor. (1991). *Development.* Sevenoaks, Kent: Hodder & Stoughton; The Open University.

(1993). Development as a robust natural process. In *Thinking about Biology*, (ed. W. D. Stein and F. Varela). SFI Studies in the Sciences of Complexity. Reading, Mass.: Addison-Wesley.

(1994a). *How the Leopard Changed Its Spots: The Evolution of Complexity*. London: Weidenfeld and Nicolson.

(1994b). Homology, development, and heredity. In *Homology: The Hierarchical Basis of Comparative Biology*, ed. B. K. Hall, pp. 230–247. San Diego: Academic Press.

(1994). Towards a science of qualities. In *The Metaphysical Foundations of Modern Science*, ed. W. Harman and J. C. Clark. Sausalito, Calif.: Institute of Noetic Sciences.

Goodwin, B. C., and Kauffman, S. A. (1990). Spatial harmonics and pattern specification in early *Drosophila* development. Part I. Bifurcation sequences and gene expression. *J. Theoret. Biol. 144*, 303–319.

(1992). Deletions and mirror symmetries in *Drosophila* segmentation mutants reveal generic properties of epigenetic mappings. In *Principles of Organization of Organisms*, (ed. J. Mittenthal and A. Baskin. Reading, Mass.: Addison-Wesley.

Goodwin, B. C., and Pateromichelakis, S. (1979). The role of electrical fields, ions, and the cortex in the morphogenesis of *Acetabularia. Planta 145*, 427–435.

Goodwin, B. C., and Saunders, P. T., editors. (1989). *Theoretical Biology: Epigenetic and Evolutionary Order from Complex Systems*. Edinburgh: Edinburgh University Press.

Goodwin, B. C., and Trainor, L. E. H. (1983). The ontogeny and phylogeny of the pentadactyl limb. In *Development and Evolution*, ed. B. C. Goodwin, N. J. Holder and C. C. Wylie, pp. 75–98. Cambridge: Cambridge University Press.

(1985). Tip and whorl morphogenesis in *Acetabularia* by calcium-regulated strain fields. *J. Theoret. Biol. 117*, 79–106.

Goodwin, B. C., Brière, C., and O'Shea, P. S. (1987). Mechanisms underlying the formation of spatial structures in cells. In *Spatial Organization in Eukaryotic Microbes*, ed. R. K. Poole and A. P. J. Trinci, pp. 1–9.

Goodwin, B. C., Kauffman, S. A., and Murray, J. D. (1993). Is morphogenesis an intrinsically robust process? *J. Theoret. Biol. 163*, 35–144.

Goodwin, B. C., Murray, J. D., and Baldwin, D. (1985). Calcium: the elusive morphogen? In "Acetabularia 1984", pp. 101–108, ed. S. Banotto, F. Cinelli, R. Billian. Belgian Nuclear Center, CEN-SCK Mol, Belgium.

Goodwin, B. C., Sibatani, A., and Webster, G. C., editors. (1989). *Dynamic Structures in Biology*. Edinburgh: Edinburgh University Press.

Goodwin, B. C., Skelton, J. C., and Kirk-Bell, S. M. (1983). Control of regeneration and morphogenesis by divalent cations in *Acetabularia mediterranea. Planta 157*, 1–7.

Gould, S. J. (1977). *Ontogeny and Phylogeny*. Cambridge, Mass.: Belknap Press.

(1982). Punctuated equilibrium – a different way of seeing. *New Scientist 94, /* 001014/141.

(1989). *Wonderful Life. The Burgess Shale and the Nature of History*. New York and London: W. W. Norton and Co.

(1991). The disparity of the Burgess Shale arthropod fauna and the limits of cladistic analysis: why we must strive to quantify morphospace. *Paleobiology 17*(4), 411–23.

Gould, S. J., and Lewontin, R. C. (1979). The spandrels of San Marco and the panglossian paradigm: a critique of the adaptationist programme. *Proc. Roy. Soc. Lond. (B). 205*, 581–98.

Govind, S., and Steward, R. (1991). Dorsoventral pattern formation in *Drosophila:* signal transduction and nuclear targeting. *Trends in Genetics 7*, 119–125.

Grainger, R. M. (1992). Embryonic lens induction: shedding light on vertebrate tissue determination. *Trends Genet. 8*, 349–355.

Green, P. B. (1987). Inheritance of pattern: analysis from phenotype to gene. *Am. Zool. 27*, 657–673.

(1989). Shoot morphogenesis, vegetative through floral, from a biophysical perspective. In *Plant Reproduction: From Floral Induction to Pollination*, ed. E. Lord and G. Bernier. *American Society of Plant Physiology* Symposium Series, vol. 1, pp. 58–75.

Green, P. B., Erickson, R. O., and Buggy, J. (1971). Metabolic and physical control of cell elongation rate. *In vitro* studies in *Nitella. Plant Physiol. 47*, 423–430.

Grene, M. (1974). Aristotle and modern biology. In *The Understanding of Nature: Essays in the Philosophy of Biology*. Dordrecht: Reidel.

Gurwitsch, A (1921) Über den Begriff des embryonalen Feldes. *W. Roux Archiv für Entwicklungsmechanik. 52*, 393–412.

Hacking, I. (1983). *Representing and Intervening: Introductory Topics in the Philosophy of Science*. Cambridge: Cambridge University Press.

Halder, G., Callaerts, P., and Gehring, W. J. (1995). Induction of ectopic eyes by targeted expression of the *eyeless* gene in *Drosophila. Science 267*, 1788–1792.

Hall, B. G. (1988). Adaptive evolution that requires multiple spontaneous mutations. *Genetics 120*, 887–897.

(1990). *Genetics 126*, 5–16.

Hall, B. K. (1991). *Evolutionary Developmental Biology*. London: Chapman and Hall.

Hammen, L. van der (1981). Type-concept, higher classification and evolution. *Acta Biotheoretica 30*, 3–48.

(1986). Some notes on taxonomic methodology. *Zool. Med. Leiden. 60*, 231–256.

Harding, K., and Levine, M. (1988). Gap genes define the limits of *antennapedia* and *bithorax* gene expression during early development in *Drosophila. EMBO J. 7(a)*, 205–214.

Harland, S. C. (1936). The genetical conception of the species. *Biol. Rev. 11*, 83–112.

Harré, R. (1970). *Principles of Scientific Thinking*. Chicago: University of Chicago Press.

(1986). *Varieties of Realism*. Oxford: Basil Blackwell.

Harré, R., and Madden, E. H. (1975). *Causal Powers. A Theory of Natural Necessity*. Oxford: Basil Blackwell.

Harrison, B. (1979). *An Introduction to the Philosophy of Language*. London: Macmillan.

Harrison, L. G., and Hillier, N. A. (1985). Quantitative control of *Acetabularia* morphogenesis by extracellular calcium: a test of kinetic theory. *J. Theoret. Biol. 114*, 177–192.

Harrison, L. G., Graham, K. T., and Lakowski, B. C. (1988). Calcium localization during *Acetabularia* whorl formation: evidence supporting a two-stage hierarchical mechanism. *Development 104*, 255–262.

Harrison, R. G. (1921). On relations of symmetry in transplanted limbs. *J. Exp. Zool. 32*, 1–136.

Hashimoto, C., Hudson, K. L., and Anderson, K. V. (1988). The *toll* gene, required for dorsal–ventral embryonic polarity, appears to encode a transmembrane protein. *Cell 52*, 629–279.

Heemskerk, J., and DiNardo, S. (1994). *Drosophila hedgehog* acts as a morphogen in cellular patterning. *Cell 76*, 449–460.

Hemmati-Brivanlou, A., and Melton, D. A. (1994). Inhibition of activin receptor signaling promotes neuralization in *Xenopus. Cell 77*, 273–281.

Hemmati-Brivanlou, A., Kelly, O. G., and Melton, D. A. (1994). Follistatin, an antagonist of activin, is expressed in the Spemann organiser and displays direct neuralizing activity. *Cell 77*, 283–295.

Hinchliffe, J. R. (1990). Towards a homology of process: evolutionary implications of experimental studies on the generation of skeletal pattern in avian limb development. In *Organisational Constraints on the Dynamics of Evolution*, ed. J. Maynard Smith and G. Vida, pp. 119–131. Manchester: Manchester University Press

(1994). Evolutionary developmental biology of the tetrapod limb. In *The Evolution of Development Mechanisms. Development* 1994 Supplement, 163–168.

Hinchliffe, J. R., and Hecht, M. (1984). Homology of the bird wing skeleton: embryological versus palaeontological evidence. *Evolutionary Biology 30*, 21–39.

Hinton, G. E., Dayan, P., Frey, B. J., and Neal, R. M. (1995). The 'sleep-wake' algorithm for unsupervised neural networks. *Science 268*, 1158–1161.

Ho., M-W. (1986). Heredity as process. *Rivista di Biologia – Biology Forum 79*, 407–447.

(1990). An exercise in rational taxonomy. *J. Theoret. Biol. 147*, 43–57.

Ho, M-W, Matheson, A., Saunders, P. T., Goodwin, B. C., and Smallcombe, A. (1987). Ether-induced segmentation defects in *Drosophila melanogaster. Roux's Arch. Dev. Biol. 196*, 511–521.

Ho, M-W., and Saunders, P. T. (1979). Beyond Neo-Darwinism: an epigenetic approach to evolution. *J. Theoret. Biol. 78*, 573–91.

Honig, L. (1984). Pattern formation during development of the amniote limb. In *The Structure, Development, and Evolution of Reptiles*, ed. M. W. J. Ferguson. Symposium 52, Zoological Society of London. London: Academic Press.

Hopfield, J. J. (1982). Neuronal networks and physical systems with emergent collective computational properties. *Proc. Nat. Acad. Sci. U.S. 81*, 3058–3062.

Hull, D. L. (1965). The effect of essentialism on taxonomy – two thousand years of stasis. *Brit. J. Phil. Sci. 15*, 314–326; *16*, 1–18.

(1967). The metaphysics of evolution. *Brit. J. Hist. Sci. 3*, 309–337

(1969). Essay review: what philosophy of biology is not. *J. Hist. Biol. 2*, 241–268.

(1975). Central subjects and historical narratives. *History and Theory 14*, 253–274.

(1976). Are species really individuals? *Systematic Zool. 25*, 174–191.

(1978). A matter of individuality. *Philosophy of Science 45*, 335–360.

(1981). Discussion: Kitts and Kitts and Caplan on species. *Philosophy of Science 48*, 141–152.

(1984). Historical entities and historical narratives. In *Minds, Machines and Evolution*, ed. C. Hookway. Cambridge: Cambridge University Press.

Hunding, A., and Engelhardt, R. (1995). Early biological morphogenesis and nonlinear dynamics. *J. Theoretical Biol. 173*, 401–413.

Hunding, A., Kauffman, S. A., and Goodwin, B. C. (1990). *Drosophila* segmentation: supercomputer simulation of prepattern hierarchy. *J. Theoret. Biol. 145*, 369–384.

Huxley, J. S., and De Beer, G. R. (1934). *The Elements of Experimental Embryology*. Cambridge: Cambridge University Press.

Huxley, J. S. (1942). *Evolution: The Modern Synthesis.* London: Allen & Unwin.

Iftode, F., Cohen, J., Ruiz, F., Torres Rueda, A., Chen-Shan, L., Adoutte, A., and Beisson, J. (1989). Development of surface pattern during division in *Paramecium.* I. Mapping of duplication and reorganisation of cortical cytoskeletal structures in the wild type. *Development 105,* 191–211.

Ingham, P. W. (1988). The molecular genetics of embryonic pattern formation in *Drosophila. Nature 335,* 25–34.

Ingham, P. W., and Martinez-Arias, A. (1992). Boundaries and fields in early embryos. *Cell 68,* 221–235.

Ingham, P. W., Pinchin, S. M., Howard, K. R., and Ish-Horowitz, D. (1985). Analysis of the *Hairy* locus in *Drosophila malanogaster. Genetics 111,* 463–486.

Jablonka, E, and Lamb, M. J. (1995). *Epigenetic Inheritance and Evolution: The Lamarckian Dimension.* Oxford: Oxford University Press.

Jacob, F. (1974). *The Logic of Living Systems.* London: Allen Laine.

Jacobson, A. G., and Sater, A. K. (1988). Features of embryonic induction. *Development 104,* 341–359.

Jaffe, L. F. (1981). The role of ionic currents in establishing developmental pattern. *Phil. Trans. Roy. Soc. B. 295,* 553–566.

Jäckle, H., Gaul, Q., and Redemann, N. (1988). Regulation and putative function of the *Drosophila* gap gene *Krüppel. Development 104* Supplement, 29–34.

Jäckle, H., Tautz, D., Schuk, R., Seifert, E., and Lehmann, R. (1986). Cross-regulatory interactions among the gap genes of *Drosophila. Nature 324,* 668–670.

Jenny, H. (1967, 1974). *Kymatic/Cymatics.* 2 vols. Basel: Basilius Presse.

Kamiya, N. (1981). Physical and chemical basis of cytoplasmic streaming. *Am. Rev. Plant Physiol. 32,* 205–236.

Kauffman, S. A. (1969). Metabolic stability and epigenesis in randomly constructed genetic nets. *J. Theoret. Biol. 22,* 437–467.

(1986). Autocatalytic sets of proteins. *J. Theoret. Biol. 119,* 1–24.

(1992). *The Origins of Order: Self-Organization and Selection in Evolution.* Oxford: Oxford University Press.

(1995). *At Home in the Universe. The Search for the Laws of Self-Organization and Complexity.* Oxford: Oxford University Press.

Kauffman, S. A., and Goodwin, B. C. (1990). Spatial harmonics and pattern specification in early *Drosophila* development. Part II. The four colour wheel model. *J. Theoret. Biol. 144,* 321–345.

Kauffman, S. A., and Levin, S. (1987). Towards a general theory of adaptive walks on rugged landscapes. *J. Theoret. Biol. 128,* 11–46.

Kent, G. C. (1969). *Comparative Anatomy of the Vertebrates.* St. Louis: C. V. Mosby.

Kitts, D. B. and Kitts, D. J. (1979). Biological species as natural kinds. *Philosophy of Science 46,* 613–622.

Knipple, D. C., Seifert, E., Rosenberg, U., Preiss, A., and Jäckle, H. (1985). Spatial and temporal patterns of *Krüppel* gene expression in *Drosophila* embryos. *Nature 317,* 40–44.

Kripke, S. A. (1980). *Naming and Necessity.* Oxford: Blackwell.

Krois, J. M. (1987). *Cassirer: Symbolic Forms and History.* New Haven, Conn.: Yale University Press.

Lacalli, T. C. (1990). Modelling the *Drosophila* pair-rule pattern by reaction-diffusion: gap input and pattern control in a 4-morphogen system. *J. Theoret. Biol. 144,* 171–194.

Laurent, M., and Fleury, A. (1995). A model with excitability and relay properties for the generation of the propagation of a Ca^{2+} morphogenetic wave in *Paramecium*.

Lawrence, P. A. (1992). *The Making of a Fly*. Oxford: Blackwell Scientific.

Le Guyader, H., and Hyver, C. (1991). Duplication of cortical units on the cortex of *Paramecium*: a model involving a Ca^{2+} wave. *J. Theoret. Biol. 150*, 261–276

Lenski, R. E., and Mittler, J. E. (1993). The directed mutation controversy and neo-Darwinism. *Science 259*, 188–194.

Levins, R., and Lewontin, R. (1985). *The Dialectical Biologist*. Cambridge, Mass.: Harvard University Press.

Lévi-Strauss, C. (1966). *The Savage Mind*. London: Weidenfeld and Nicholson.
(1968). History and Anthropology. In *Structural Anthropology I*. London: Allen Lane, Penguin Press.
(1969). *Totemism*. Harmondsworth: Penguin Books

Lewis, J. H., and Wolpert, L. (1976). The principle of nonequivalence in development. *J. Theoret. Biol. 62*, 479–490.

Lovejoy, A. O. (1968). Recent criticism of the Darwinian theory of recapitulation: its grounds and its initiator. In *Forerunners of Darwin: 1745–1859*, ed. B. Glass, O. Temkin and W. L. Strauss. Baltimore: Johns Hopkins University Press.

Lovelock, J. (1979). *A New Look at Life on Earth*. Oxford: Oxford University Press.
(1991). *Gaia: The Practical Science of Planetary Medicine*. London: Gaia Books.

Løvtrup, S. (1984). Ontogeny and Phylogeny. In *Beyond Neo-Darwinism*. ed. M.-W. Ho and P. Saunders. London: Academic Press.

MacCabe, J. A., Saunders, J. W., and Picket, M. (1973). The control of the antero-posterior and dorsoventral axes in embryonic chick limbs constructed of dissociated and reaggregated limb-bud mesoderm. *Dev. Biol. 31*, 323–335.

Macdonald, P. M., and Struhl, G. (1986). A molecular gradient in early *Drosophila* embryos and its role in specifying body pattern. *Nature, Lond. 324*, 537–545.

Macdonald, P. M., Ingham, R., and Struhl, G. (1986). Isolation, structure, and expression of *even-skipped*: a second pair-rule gene of *Drosophila* containing a homeobox. *Cell 47*, 721–734.

Maden, M. (1982a). Vitamin A and pattern formation in the regenerating limb. *Nature 295*, 672–675.
(1982). Supernumerary limbs in amphibians. *Am. Zool. 22*, 131–142.
(1983). The effect of vitamin A on the regenerating axolotl limb. *J. Embryol. Exp. Morphol. 77*, 273–295.

Maden, M., and Keeble, S. (1987). The role of cartilage and fibronectin during respecification of pattern induced in the regenerating amphibian limb by retinoic acid. *Differentiation 36*, 175–184.

Malacinski, G. M. (1990). *Cytoplasmic Organization Systems*. New York: McGraw-Hill.

Mandelbrot, B. B. (1977). *Fractals: Form, Chance, and Dimension*. San Francisco: W. H. Freeman.

Matheson, A. (1991). Teratology and the development of pattern. Unpublished Ph.D. Thesis, Open University.

Maturana, H. R., and Varela, F. J. (1987). *The Tree of Knowledge. The Biological Roots of Human Understanding*. Boston and London: New Science Library, Shambala.

May, R. M. (1976). Simple mathematical models with very complicated dynamics. *Nature 261*, 459–467.

Maynard Smith, J., and Szathmary, E. (1995). *The Major Transitions in Evolution.* Oxford: W. H. Freeman.

Mayr, E. (1959). Agassiz, Darwin and evolution. *Harvard Library Bulletin 13,* 165–194.

——— (1963). *Animal Species and Evolution.* London: Oxford University Press.

——— (1966). Introduction. In C. Darwin, *On the Origin of Species.* Reprinted. Cambridge Mass.: Harvard University Press.

——— (1969). *Principles of Systematic Zoology.* New York: McGraw-Hill.

——— (1976). *Evolution and the Diversity of Life.* Cambridge, Mass.: Harvard University Press.

——— (1988). *Toward a New Philosophy of Biology.* Cambridge, Mass.: Harvard University Press.

Meinhardt. H. (1982). *Models of Biological Pattern Formation.* London: Academic Press.

——— (1986). Hierarchical inductions of cell states – a model for segmentation in *Drosophila. J. Cell Sci. Suppl. 4,* 357–381.

Menzel, D. (1992). *The Cytoskeleton of the Algae.* Boca Raton, Fla.: CRC Press.

Menzel, D., and Elsner-Menzel, C. (1989). Induction of actin-based contraction in the siphonous green alga *Acetabularia* (Chlorophyceae) by locally restricted calcium influx. *Bot. Acta 102,* 164–171.

Mepham, J. (1972). The structuralist sciences and philosophy. In *Structuralism: An Introduction,* ed. D. Robey. Oxford: Oxford University Press.

Minelli, A. (1993). *Biological Systematics: The State of the Art.* London: Chapman and Hall.

Miramontes, O., Sole, R., and Goodwin, B. C. (1993) Collective behaviour of random-activated mobile cellular automata. *Physica D 63,* 145–160.

Mittenthal, J. E., and Beloussov, L. V. (1991). Processes that may shape surfaces in embryos. Technical Report CCSR-91-7, Centre for Complex Systems Research, The Beckman Institute, University of Illinois.

Mjolsnes, E., Garrett, C. D., Reinitz, J., and Sharp, D. H. (1995). Modeling the connection between development and evolution: preliminary report. In *Evolution and Biocomputation: Computational Models of Evolution,* ed. F. H. Eeckman and W. Banzhaf, pp 103–122. Berlin: Springer Verlag.

Monod, J. (1972). *Chance and Necessity.* London: Collins.

Morgan, B. A., and Tabin, C. J. (1994). Hox genes and growth: early and late roles in limb bud morphogenesis. In *The Evolution of Developmental Mechanisms. Development* 1994 Supplement, 181–186.

Morgan, B. A., Izpisua-Belmonte, J. C., Duboule, D., and Tabin, C. J. (1992). Ectopic expression of Hox4.6 in the avian limb bud causes homeotic transformation of anterior structures. *Nature 358,* 236–239.

Morgan, T. H. (1903). *Evolution and Adaptation.* New York: Macmillan.

Murray, J. D. (1989). *Mathematical Biology.* Berlin: Springer-Verlag.

Murray, J. D., and Oster, G. (1984). Generation of biological pattern and form. *IMA J. Maths in Med. and Biol. 1,* 51–75.

Nagorcka, B. N. (1988). A pattern formation mechanism to control spatial organization in the embryo of *Drosophila melanogaster. J. Theoret. Biol. 132,* 277–306.

Needham, J. (1942). *Biochemistry and Morphogenesis.* Cambridge: Cambridge University Press.

Needham, A. E. (1952). *Regeneration and Wound Healing.* London: Macmillan.

Nelson, G. (1978). Ontogeny, phylogeny, paleontology and the biogenetic law. *Syst. Zool. 27,* 324–345.

Newman, S. A., and Comper, W. D. (1990). 'Generic' physical mechanisms of morphogenesis and pattern formation. *Development 110,* 1–18.

Newman, S. A., and Frisch, H. L. (1979). Dynamics of skeletal pattern formation in the developing chick limb. *Science 205,* 662–668.

Ng, S. F., and Frankel, J. (1977). 180° rotation of ciliary rows and its morphogenetic implications in *Tetrahymena pyriformis. Proc. Nat. Acad. Sci. USA 74,* 1115–1119.

Nicolis, G., and Prigogine, I. (1987). *An Introduction to Complexity.* New York: W. H. Freeman.

Nossal, R. (1988). On the elasticity of cytoskeletal networks. *Biophys. J. 53,* 349–359.

Nüsslein-Volhard, C. (1977). Genetic analysis of pattern formation in the embryo of *Drosophila melanogaster.* Characterisation of the maternal effect mutant Bicaudal. *Roux's Arch. 183,* 244–268.

(1979). Maternal effect mutations that alter the spatial coordinates of the embryo. In *Determinants of Spatial Organization,* ed. S. Subtelny and I. R. Konigsberg, pp. 185–211. New York: Academic Press.

Nüsslein-Volhard, C., and Wieschaus, E. (1980). Mutations affecting segment member and polarity in *Drosophila. Nature 287,* 795–801.

O'Shea, P., Goodwin, B., and Ridge, I. (1990). A vibrating electrode analysis of extracellular ion currents in *Acetabularia acetabulum. J. Cell Science 97,* 505–508.

Odell, G., Oster, G. F., Burnside, B., and Alberch, P. (1981). The mechanical basis of morphogenesis. *Devel. Biol. 85,* 446–462.

Oosawa, F., Kasai, M., Hatano, S., and Asakura, S. (1966). In *Principles of biomolecular organisation,* ed. G. E. W. Wolstenholme and M. O'Connor, pp. 273–303. Boston,: Little, Brown.

Osborn, H. F. (1915). On the origin of single characters as observed in fossil and living animals and plants. *Amer. Nat. 49,* 193–239.

Oster, G., and Alberch, P. (1982). Evolution and bifurcation of developmental programs. *Evolution 36,* 444–459.

Oster, G. F., and Odell, G. M. (1983). The mechanochemistry of cytogels. In *Fronts, Interfaces and Patterns,* ed. A. Bishop. Amsterdam: North Holland, Elsevier Science Division.

Oster, G. F., Murray, J. D., and Harris, A. (1983). Mechanical aspects of mesenchymal morphogenesis. *J. Embryol. Exp. Morph. 78,* 83–125.

Oster, G. F., Murray, J. D., and Maini, P. K. (1985). A model for chondrogenic condensations in the developing limb: the role of the extracellular matrix and cell fractions. *J. Embryol. Exp. Morphol. 89,* 93–112.

Oster, G. F., Shubin, N., Murray, J. D., and Alberch, P. (1988). Evolution and morphogenetic rules: the shape of the vertebrate limb in ontogeny and phylogeny. *Evolution 42,* 862–884.

Ouweneel, W. J. (1976). Developmental genetics of homoeosis. *Advances in Genetics 18,* 179–236.

Owen, R. (1848). *Report on the Archetype and Homologies of the Vertebrate Skeleton.* London: Voorst.

Oyama, S. (1985). *The Ontogeny of Information: Developmental Systems and Evolution.* Cambridge: Cambridge University Press.

(1989). Ontogeny and the central dogma: do developmentalists need the concept of genetic programming in order to have an evolutionary perspective? In *Minnesota Symposium on Child Psychology,* vol. 22, ed. M. Gunnar.

Patel, N. H. (1994). Developmental evolution: insights from studies of insect segmentation. *Science 266,* 581–590.

Pateman, T. (1987). Philosophy of linguistics. In *New Horizons in Linguistics*. ed. R. Coates, M. Deuchar, G. Gazdar and J. Lyons. Harmondsworth: Penguin Books.

Pautou, M. P. (1973). Analyse de la morphogénèse du pied des Oiseaux à l'aide mélanges cellulaires interspécifiques. I. Etude morphologique. *J. Embryol. Exp. Morph. 69*, 1–6.

Piaget, J. (1971). *Structuralism*. London: Routledge & Kegan Paul.

Pichot, A. (1994) Definition and identity of the living being. *World Futures 42*, 21–29. Amsterdam: OPA.

Popper, K. R. (1966). *The Open Society and Its Enemies*, vol. 2. 5th edn. London: Routledge & Kegan Paul.

Putnam, H. (1975). Is semantics possible? and The meaning of meaning. In *Mind, Language and Reality: Philosophical Papers*, vol. 2. Cambridge: Cambridge University Press.

Quiring, R. Walldorf, U., Kloter, U., and Gehring, W. J. (1994). Homology of the *eyeless* gene of *Drosophila* to the *small eye* gene in mice and *aniridia* in humans. *Science 265*, 785.

Reid, R. G. B. (1985). *Evolutionary Theory: The Unfinished Synthesis*. London: Croom Helm

Reinitz, J., and Levine, M. (1990). Control of the initiation of homeotic genes by the gap genes *giant* and *tailless* in *Drosophila*. *Dev. Biol. 140*, 57–72.

Reinitz, J., and Sharp, D. H. (1995). Mechanism of *eve* stripe formation. Mechanisms of Devel. 133–158.

Reinitz, J., Mjolsnes, E., and Sharp, D. H. (1995) Model for cooperative control of positional information in *Drosophila* by *Bicoid* and maternal *hunchback*. *J. Exp. Zool. 271*, 47–56.

Rosenberg, S. M., Longerich, S. Gee, P. and Harris, R. S. (1994). Adaptive mutation by deletions in small mononucleotide repeats. *Science 265*, 405–408.

Roux, W. (1905). *Die Entwicklungsmechanik, ein neur Zweig der biologischen Wissenschaft; Vorträge und Aufsätze über Entwicklungsmechanik der Organismen*, vol. I. Leipzig: Wilhelm Engelmann.

Russell, E. (1916). *Form and Function*. London: John Murray.

Russell, M. (1985) Positional information in insect segments. *Devel. Biol. 108*, 269–283.

Saha, M. S., Spann, C. L., and Grainger, R. M. (1989). Embryonic lens induction: more than meets the optic vesicle. *Cell Diff. Devel. 28*, 153–72.

Salvini-Plawen, L. V., and Mayr, E. (1977). On the evolution of photoreceptors and eyes. *Evolutionary Biology 10*, 207–263.

Sander, K. (1983). The evolution of patterning mechanisms: gleanings from insect embryogenesis and spermatogenesis. In *Development and Evolution*, ed. B. Goodwin, N. Holder, and C. Wylie, pp 137–159.

(1986). The role of genes in ontogenesis: evolving concepts from 1883 to 1983 as perceived by an insect embryologist. In *A History of Embryology*, ed. T. J. Horder, J. A. Witkowski and C. C. Wylie, pp. 363–395. Cambridge: Cambridge University Press.

Saunders, J. W., Cairns, J. M. & Gasseling, M. T. (1957). The role of the apical ectodermal ridge of ectoderm in the differentiation of the morphological structure and inductive specificity of limb parts in the chick. *J. Morphol. 101*, 57.

Shubin, N. H. (1994). History, ontogeny, and evolution of the archetype. In *Homology: The Hierarchical Basis of Comparative Biology*, ed. B. K. Hall, pp. 230–247. San Diego: Academic Press.

Shubin, N. H., and Alberch, P. (1986). A morphogenetic approach to the origin and basic organization of the tetrapod limb. *Evolutionary Biology 20*, 319–387.

Sibatani, A. (1985). Molecular biology: a structuralist revolution. *Riv. Biol. 78*, 373–97.

Skarda, C. A., and Freeman, W. F. (1989). How brains make chaos in order to make sense of the world. *Behav. and Brain Sci. 10*, 161–195.

Slack, J. M. W. (1976). Determination of polarity in the limb. *Nature 261*, 44–46.

Slack, J. M. W., Holland, P. W. H., and Graham, C. F. (1993). The zootype and the phylotypic stage. *Nature 361*, 490–492.

Smith, J. C. (1994). *Hedgehog*, the floor plate, and the zone of polarizing activity. *Cell 76*, 193–196.

Sober, E. (1993). *Philosophy of Biology*. Oxford: Oxford University Press.

Sonneborn, T. M. (1970). Gene action in development. *Proc. Roy. Soc. Lond. B 176*, 347–366.

Spemann, H. (1938). *Embryonic Development and Induction*. New Haven, Conn.: Yale University Press.

Stein, W. D., and Bronner, F. (1989). *Cell Shape; Determinants, Regulation, and Regulatory Role*. San Diego: Academic Press.

Stern, C. (1954). Two or three bristles? *Am. Scientist 42*, 213–247.

Stewart, I. (1989). *Does God Play Dice?* London: Penguin.

Stewart, I., and Golubitsky, M. (1992). *Fearful Symmetry: Is God a Geometer?* Oxford: Basil Blackwell.

Struhl, G., Johnston, P., and Lawrence, P. A. (1992). Control of *Drosophila* body pattern by the *hunchback* morphogen gradient. *Cell 69*, 237–249.

Summerbell, D. (1981). The control of growth and the development of pattern across the anteroposterior axis of the chick limb. *J. Embryol. Exp. Morphol. 63*, 161–180.

Swift, J. (1726). *Gulliver's Travels*. Reprinted, Harmondsworth: Penguin Books, 1967.

Symonds, N. (1994). Directed mutation – a current perspective. *J. Theoret. Biol., 169*, 317–322.

Tabin, C. J. (1992). Why we have (only) five fingers per hand: Hox genes and the evolution of paired limbs. *Development 116*, 289–296.

Taylor, C. (1975). *Hegel*. Cambridge: Cambridge University Press.

Thompson, D'Arcy W. (1942). *On Growth and Form*. 2nd edn. Cambridge: Cambridge University Press.

Thorogood, P. V. (1991). The development of the teleost fin and implications for our understanding of tetrapod limb evolution. In *Developmental Patterning of the Vertebrate Limb*, ed. J. R. Hinchliffe, L. M. Hurle and D. Summerbell, pp. 347–354.

Tickle, C., Alberts, B., Wolpert, L., and Lee, J. (1982). Local application of retinoic acid to the limb bud mimics the action of the polarizing zone. *Nature 296*, 564–566.

Turing, A. M. (1952). The chemical basis of morphogenesis. *Phil. Trans. Roy. Soc. B 237*, 37–72.

Van der Meer, J. M. (1984). Parameters influencing reversal of segment sequences in posterior egg fragments of *Callosobruchus (Coleoptera)*. *Wilhelm Roux Arch. Dev. Biol. 193*, 339–356.

Waddington, C. H. (1940). *Organisers and Genes*. Cambridge: Cambridge University Press.

(1956). *Principles of Embryology*. London: George Allen & Unwin

(1957). *The Strategy of the Genes*. London: Allen & Unwin;

(1962). *New Patterns in Genetics and Development.* New York: Columbia University Press.

Wagner, G. P. (1989). The origin of morphological characters and the biological basis of homology. *Evolution 43,* 1157–1171.

(1994). Homology and the mechanisms of development. In *Homology: The Hierarchical Basis of Comparative Biology,* ed B. K. Hall, pp 274–299. San Diego: Academic Press.

Webster, G. (1971). Morphogenesis and pattern formation in hydroids. *Biol. Rev. 46,* 1–46.

(1992). William Bateson and the science of form. In W. Bateson, *Materials for the Study of Variation.* Reprinted 1992. Baltimore: Johns Hopkins University Press.

Webster, G. (1984). The relations of natural forms. In *Beyond Neo-Darwinism,* ed. M.-W. Ho and P. T. Saunders, pp. 193–218. London: Academic Press.

Webster, G., and Goodwin, B. C. (1982). The origin of species: a structuralist approach. *J. Soc. Biol. Struct. 5,* 15–47.

Webster, G. and Goodwin, B. C. (1988). *Il Problema della Forma in Biologia.* Rome: Armando Editore.

Weir, M. P., and Kornberg, T. (1985). Patterns of *engrailed* and *fushi tarazu* transcripts reveal novel intermediate stages in *Drosophila* segmentation. *Nature 318,* 433–445.

Weismann, A. (1883). On Heredity. Reprinted in *A Sourcebook in Animal Biology,* ed. T. S. Hall. New York: Hafner.

(1885). The continuity of the germ plasm as the foundation of a theory of heredity. Reprinted in *Readings in Heredity and Development,* ed. J. A. Moore. New York: Oxford University Press, 1972.

Weiss, P. (1939). *Principles of Development.* New York: Holt.

Weliky, M., Minsuk, S., Keller, R, and Oster, G. (1991). Notochord morphogenesis in *Xenopus laevis:* simulation of cell behaviour underlying tissue convergence and extension. *Development 113,* 1231–1244.

Wilby, O. K., and Ede, D. A. (1976). Computer simulation of vertebrate limb development. In *Automata, Languages, Development,* ed. B. Lindenmeyer and G. Rozenberg, pp. 143–61. Amsterdam: North-Holland.

Willmer, E. N. (1960). *Cytology and Evolution.* New York: Academic Press.

Winfree, A. T. (1980). *The Geometry of Biological Time.* New York: Springer-Verlag.

(1984). A continuity principle for regulation. In *Pattern Formation,* ed. G. M. Malacinski and S. V. Bryant, pp. 103–124. London: Macmillan.

(1987). *When Time Breaks Down.* Princeton, N.J.: Princeton University Press.

Wolpert, L. (1969). Positional information and the spatial pattern of cellular differentiation. *J. Theoret. Biol. 25,* 1–47.

(1971). Positional information and pattern formation. *Curr. Top. Devel. Biol. 6,* 183–224.

(1991). *The Triumph of the Embryo.* Oxford: Oxford University Press.

Wolpert, L., and Lewis, J. (1975). Towards a theory of development. *Fed. Proc. 34,* 14–20.

Wolpert, L., and Stein, W. D. (1984). Positional information and pattern formation. In *Pattern Formation,* ed. G. M. Malacinski and S. V. Bryant. London: Macmillan.

Woodger, J. H. (1930). The "concept of organism" and the relation between embryology and genetics. *Quart. Rev. Biol. 5,* 1–22.

(1945). On biological transformations. In *Essays on Growth and Form Presented to D'Arcy Wentworth Thompson,* ed. W. E. Le Gros Clark and P. B. Medawar. Oxford: Oxford University Press.

Woodger, J. H. (1948). Observations on the present state of embryology. *Symp. Soc. Exp. Biol. 2,* 345–360.

Wright, G. H. von. (1971). *Explanation and Understanding.* Ithaca, N.Y.: Cornell University Press.

(1974). *Causality and Determinism.* New York: Columbia University Press.

Zwilling, E. (1964). Development of fragmented and dissociated limb bud mesoderm. *Dev. Biol. 9,* 20–37.

Index

on morphogenesis, 94, 95, 98, 122
on transformation, 118–119, 120,
121
Wright, G. H. von, 51, 74

X

Xenopus, activin/follistatin role in
development of, 185

Z

zone of polarising activity (ZPA), 135,
139, 143
role in limb development, 146, 147,
152
zootype, 244, 248
description of, 187
zygote structure, theory of, 121